Gérard Rossé (Ed.)

Supercritical Fluid Chromatography

Also of Interest

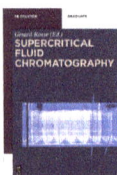

Supercritical Fluid Chromatography.
Volume 2
Rossé (Ed.), 2018
ISBN 978-3-11-061893-8, e-ISBN 978-3-11-061898-3

Electrophoresis.
Theory and Practice
Michov, 2019
ISBN 978-3-11-033071-7, e-ISBN 978-3-11-033075-5

Inorganic Trace Analytics.
Trace Element Analysis and Speciation
Matusiewicz, Bulska (Eds.), 2017
ISBN 978-3-11-037194-9, e-ISBN 978-3-11-036673-0

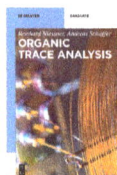

Organic Trace Analysis.
Nießner, Schäffer, 2017
ISBN 978-3-11-044114-7, e-ISBN 978-3-11-044115-4

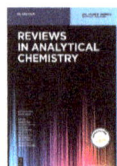

Reviews in Analytical Chemistry.
Editor-in-Chief: Israel Schechter
e-ISSN 2191-0189

Supercritical Fluid Chromatography

Volume 1

Edited by
Gérard Rossé

DE GRUYTER

Editor
Dr. Gérard Rossé
Dart NeuroScience LLC
12278 Scripps Summit Dr
San Diego, CA, 92131
USA
Department of Pharmacology and Physiology
College of Medicine, Drexel University
New College Building, 245 North 15th Street
Philadelphia, PA 19102
USA
gerardcrosse@gmail.com

ISBN 978-3-11-050075-2
e-ISBN (PDF) 978-3-11-050077-6
e-ISBN (EPUB) 978-3-11-049812-7

Library of Congress Control Number: 2018031531

Bibliographic information published by the Deutsche Nationalbibliothek
The Deutsche Nationalbibliothek lists this publication in the Deutsche Nationalbibliografie; detailed bibliographic data are available on the Internet at http://dnb.dnb.de.

© 2019 Walter de Gruyter GmbH, Berlin/Boston
Typesetting: Integra Software Services Pvt. Ltd.
Printing and binding: CPI books GmbH, Leck
Cover image: sumos / iStock / Getty Images Plus

www.degruyter.com

Preface

When Oleg Lebedev from De Gruyter reached out to me about writing a book on supercritical fluid chromatography (SFC), I immediately replied "Yes." In the last few years, the field of SFC has seen developing rapidly and a special edition on SFC appears to be particularly timely. I am most grateful to not only the pioneers in the field of SFC but also the renowned experts from industry and academia for contributing 15 chapters and allowing us to assemble a collection of truly exciting ideas and reports. I am equally grateful to my colleagues and friends who agreed to act as reviewers. To reveal the broad impact of modern SFC in the life sciences, I decided to cover applications of the technique in the pharmaceutical industry, the food industry, the fragrance and perfume industry, natural products, and substance abuse (doping). The 15 chapters are divided into two volumes dedicated to the concepts, potentials, and limitations of the relevant SFC applications. In some cases, I retained some redundancy of content to represent the authors' individual views on a topic. Each volume has a balanced number of chapters and can be read independently. This two-volume series opens with an overview of the history and expectant future of SFC and continues with recent applications in the pharmaceutical industry and other fascinating areas of science. The two volumes are not a comprehensive treatise on the subject, nor are they a repackaging of what others have pioneered and written. Their intention is to serve as a source of inspiration and stimulation for readers to continue exploring the possibilities of chromatography and synthesis with a supercritical fluid.

For me, a simple way to describe SFC is "today's technology enabling tomorrow's innovation." SFC is not new but went through different phases of acceptance, development, and success. In the 1980s it was considered science-fiction technology and received a burst of interest in the 1990s. In 2010, the efforts of two major manufacturers in developing analytical instruments to meet industry standards reawakened the scientific community. SFC bridges the gap between gas chromatography (GC) and liquid chromatography (LC), and its broader impact on numerous frontier areas of industrial and environmental analytical chemistry is being studied. The successful coupling of SFC to mass spectrometers (SFC-MS) is now straightforward and provides highly reliable and robust SFC-MS instruments with applications in the sample analysis and mass-triggered purification of complex mixtures. In the last few years, multiple studies have demonstrated SFC/SFC-MS to be as robust, reliable, and precise as LC/LC-MS. Scientists should see modern SFC as another form of liquid chromatography that happens to use CO_2 as the mobile phase. The main advantages of SFC are compelling. It is faster than LC, it significantly reduces solvent consumption and waste. For sample, purification SFC fraction dry-down time is 15-fold faster than for aqueous fractions obtained in LC. SFC is also a "green" technology using CO_2 produced in existing fermentation plants, and it limits or eliminates the use of most toxic organic solvents.

https://doi.org/10.1515/9783110500776-201

SFC has found its place in the pharmaceutical industry with an increasing body of applications for chiral and achiral molecules in both the research and development phases of the drug discovery process. As illustrated in this two-volume series, the current interest in SFC extends well beyond the pharmaceutical industry. Chapters encompassing applications for polar and nonpolar mixtures of importance are covering widely disparate areas in substance abuse, natural products including cannabinoids, bioactive lipids, flavor, and fragrance. At a time dominated by aggressive project timelines, the need to deliver products faster and in a world seeking renewable technologies for solving the waste and disposal problems, SFC can contribute to the more expedient delivery of better compounds while minimizing the impact on the environment in which we live.

The initiation of this book coincided with my involvement with Dart NeuroScience in building and using one of the largest operation-based SFC-MS systems in the pharmaceutical industry. During this time I had the opportunity to measure the impact of SFC-MS on productivity and cost reduction. The closure of our R&D operations led me to reflect on the achievements and value of SFC. SFC is truly enabling future innovations and could revolutionize the field of chromatography. However, the full potential of SFC has yet to be realized. Understanding the fundamentals of SFC, setting even higher requirements on system quality and cost-effectiveness, and building innovative systems will continue to elevate interest, expand the breadth of applications, and grow the market for SFC-/SFC-MS-based technologies.

I want to thank Oleg Lebedev, Lena Stoll, and Sabina Dabrowski from De Gruyter for supporting me in the editing process. I dedicate this book to the Dart NeuroScience family. Special thanks go to the team working on implementing SFC-MS with me and to Tim Tully, company founder, who enable me to work in this fascinating area of science. Steven de Belle did a great job in providing advice and reviewing my grammars. John van Antwerp, Ronan Cleary, and many great friends at Waters Corporation helped solve challenges and break new frontiers. I am very grateful to my wife, Frédérique, and my beautiful daughters, Alyssa and Noanne.

San Diego, CA
September 2018   Gérard Rossé

Contents

List of Contributors

Cristina Anta
Eli Lilly & Company
Avenida de la Industria 30
28108 Alcobendas, Madrid
Spain
anta_cristina@lilly.com

Chandan L. Barhate
Discovery Chemistry and Technology
AbbVie Inc.
1 North Waukegan Rd
Dept. R467, Bldg. AP9
North Chicago, IL 60064–6114
USA

Matthew Belvo
Eli Lilly & Company
Indianapolis IN 46285
USA

Terry A. Berger
SFC Solutions, Inc.
9435 Downing St.
Englewood, FL 34224
tabergersfc@aol.com

John Burnett
Eli Lilly & Company
Erl Wood Manor
Sunninghill Rd
Windlesham, Surrey G20 6PH
United Kingdom

Thomas Castle
Eli Lilly & Company
Indianapolis, IN 46285
USA
castle_thomas_m@lilly.com

Steven M. Collier
Waters Corporation
34 Maple Street
Milford, MA 01757
USA
Steven_M_Collier@waters.com

Vincent Desfontaine
School of Pharmaceutical Sciences
University of Geneva
University of Lausanne
Rue Michel Servet 1, CMU
1211 Geneva 4
Switzerland

María Luz de la Puente
Eli Lilly & Company
Avenida de la Industria 30
28108 Alcobendas, Madrid
Spain
de_la_puente_maria_luz@lilly.com

Eric R. Francotte
FrancotteConsulting Orisstrasse 4
4412 Nuglar
Switzerland
eric.francotte@francotteConsulting.ch

Davy Guillarme
School of Pharmaceutical Sciences
University of Geneva
University of Lausanne
Rue Michel Servet 1, CMU
1211 Geneva 4
Switzerland

Jason F. Hill
Waters Corporation
34 Maple Street
Milford, MA 01757
USA
Jason_F_Hill@waters.com

Erin E. Jordan
Discovery Chemistry and Technology
AbbVie Inc.
1 North Waukegan Rd
Dept. R467, Bldg. AP9
North Chicago, IL 60064–6114
USA

https://doi.org/10.1515/9783110500776-202

Andreas Kaerner
Eli Lilly & Company
Indianapolis, IN 46285
USA
kaerner_andreas@lilly.com

Tiia Kuuranne
Swiss Laboratory for Doping Analyses
University Center of Legal Medicine
Lausanne-Geneva
Centre Hospitalier Universitaire Vaudois
University of Lausanne
Chemin des Croisettes 22
1066 Epalinges
Switzerland

Eric Lesellier
University of Orléans
Institute of Organic and Analytical Chemistry
BP 6759
45067 Orléans cedex 2
France
eric.lesellier@univ-orleans.fr

Pilar Lopez
Eli Lilly & Company
Avenida de la Industria 30
28108 Alcobendas, Madrid
Spain
lopez-soto_pilar@lilly.com

Alexander Marziale
Novartis Pharma AG
Novartis Institutes for Biomedical Research (NIBR)
Klybeckstrasse 141
4057 Basel
Switzerland
alexander.marziale@novartis.com

Raul Nicoli
Swiss Laboratory for Doping Analyses
University Center of Legal Medicine
Lausanne-Geneva
Centre Hospitalier Universitaire Vaudois
University of Lausanne
Chemin des Croisettes 22
1066 Epalinges
Switzerland

Arancha Sonia Marin
Eli Lilly & Company
Avenida de la Industria 30
28108 Alcobendas, Madrid
Spain
marin_aranzazu@lilly.com

Thomas Perun
Eli Lilly & Company
Avenida de la Industria 30
28108 Alcobendas, Madrid
Spain

Alfonso Rivera
Eli Lilly & Company
Avenida de la Industria 30
28108 Alcobendas, Madrid
Spain

Philip A. Searle
Discovery Chemistry and Technology
AbbVie Inc.
1 North Waukegan Rd
Dept. R467, Bldg. AP9
North Chicago, IL 60064–6114
USA
Philip.Searle@abbvie.com

Eric Seest
Eli Lilly & Company
Indianapolis, IN 46285
USA

Abhijit Tarafder
Waters Corporation
34 Maple Street
Milford, MA 01757
USA
Abhijit_Tarafder@waters.com

Jean-Luc Veuthey
School of Pharmaceutical Sciences
University of Geneva
University of Lausanne
Rue Michel Servet 1, CMU
1211 Geneva 4
Switzerland

Caroline West
University of Orléans
Institute of Organic and Analytical
Chemistry
BP 6759
45067 Orléans cedex 2
France
caroline.west@univ-orleans.fr

Craig White
Eli Lilly & Company
Erl Wood Manor
Sunninghill Rd
Windlesham, Surrey G20 6PH
United Kingdom
white_craig_t@lilly.com

Terry A. Berger

1 A history of supercritical fluid chromatography (SFC)

Abstract: Supercritical fluid chromatography (SFC) has slowly developed over more than 60 years, and has only fairly recently become mainstream. Some of the reasons for such slow development were a few shaky concepts from the nineteenth century and a lack of adequate technology. When SFC was first proposed in 1957, high-performance liquid chromatography (HPLC) did not exist. Gas chromatography (GC) was in its infancy but growing rapidly. SFC was seen as a possible means to extend GC to heavier, low-volatility compounds. The details regarding fluids were not completely understood. Pumping highly compressible fluids at high pressures was fraught with difficulties. All equipment used up until the early 1980s were homemade and surprisingly crude. Most of the work of the first 20 years pointed in a direction different from where we wound up. Many of the scientists involved were some of the biggest names in the field of chromatography. The stature of these scientists made it difficult to change directions to pursue a different path later on. Consequently, the 1980s were filled with controversies. The viewpoint expressed in this chapter is somewhat biased since the author was directly involved during the years of the greatest controversies.

Nowadays, instrumentation for SFC is as robust as HPLC. Most applications use small particle columns and binary mobile phases, often with gradient elution. There is accurate independent dynamic control of flow, composition, temperature, and outlet pressure. The physical and chemical nature of the mobile phases is reasonably well understood. None of this was true for the first 25+ years of SFC.

Keywords: pre-1980 confusion, 1980's controversies, Modern SFC, mainstream chiral analysis, 2nd and 3rd generations

1.1 The beginnings up to 1980

The existence of a critical point was first discovered by Charles Cagniard de la Tour in 1822 [1]. He showed that CO_2 could be liquefied at temperatures up to 31 °C by increasing pressure up to 73 bar, but above 31 °C, it could not be liquefied even at 3,000 bar. Dmitri Mendeleev may have named it the "critical point" in 1860. In 1869, Thomas Andrews [2] also described the end of the liquid/vapor equilibrium line as the "critical point" and coined the terms "critical temperature" and "critical pressure" using CO_2.

Terry A. Berger, SFC Solutions, Inc.,9435 Downing St., Englewood, FL 34224

https://doi.org/10.1515/9783110500776-001

Ten years later, in a short communication, Hannay and Hogarth [3] described the solubility of a number of inorganic salts and chlorophyll in several fluids above their critical point. Most of the work was performed with supercritical methanol or ethanol. According to Krukonis [4], this report was greeted with disbelief, which was only overcome with substantially more work. The criticisms were quite severe, and the response remains the same. In a following paper, Hannay stoutly defended his previous work and provided substantial further evidence. He claimed that "The gas must have a certain density before it will act as a solvent." This was later called "threshold density" by Giddings. Hannay [5] has been widely quoted as having said: "The liquid condition of fluids has very little to do with their solvent power, but only indicates molecular closeness. Should this closeness be attained by external pressure instead of internal attraction, the result is the same or even greater solvent power...." He ignored or was unaware of the differences in intermolecular interactions by different types of solvents and suggesting it is only density that is important. Such a claim is, of course, unrealistic. Today, beside dispersive interactions, we also look at proton donor, proton acceptor, and dipole and electrostatic interactions.

There is no indication that anyone up to this point knew what the density of the various fluids was above their critical points. Perhaps more importantly, there also appears to have been no attempt to try similar high-pressure experiments, but below the critical temperature, using fluids such as CO_2, which are gasses under atmospheric conditions. Typical of developments with supercritical fluids, it was only in 1906 (27 years later) that Buchner [6] studied the solubility of naphthalene in supercritical CO_2. Naphthalene became a model compound for various studies.

After another 50+ years, Jim Lovelock, the designer of the gas chromatography (GC) that analyzed the Marian atmosphere, inventor of the helium and argon ionization detectors, and the electron capture detector, suggested, at a 1956 international meeting, that some gases compressed above their critical point might be usable as a mobile phase in GC, to elute larger compounds with little or no volatility. It is unclear if he knew of Hannay's work. He typed a one-page letter he had notarized in 1957, copies of which still exist. He suggested the name "critical state chromatography." Lovelock never did any lab work to test his supposition.

Lovelock suggested that some inorganic gases such as SO_2 under high pressure were likely to solvate more polar molecules that were not volatile in GC. He also suggested that columns "like Desty used" would be useful. Golay [7] had presented his theory for open tubular columns for GC, at the spring meeting of the American Chemical Society in 1956, and subsequently published in 1957. Dennis Desty [8] was an early leader in their application using very small bore glass and metal columns. Years later this letter resulted in an argument about the patentability of capillary SFC, as proposed by Novotny et al. [9], but we are getting ahead of the story.

1.1.1 Klesper and Corwin

It took another 5 years after Lovelock's suggestion, before Klesper et al. [10] published the first report of what we now call supercritical fluid chromatography in 1962. Of all the places, they chose to report the results as a brief (approximately two pages) preliminary "Communication to the Editor" in the *Journal of Organic Chemistry*. They used dichlorodifluoromethane and chlorodifluoromethane (refrigerants) as the mobile phases to separate metal porphyrins of the types found in crude oil. These porphyrins were thought to offer a means of understanding the origins of petroleum but were too thermally unstable for regular GC.

Details were sketchy. There was no pump. In fact there were no pumps that could accurately control the flow rate of a compressible fluid at high pressures at that time. The fluid was placed in a closed reservoir, which was then heated in an oven to generate higher pressure, and then passed through a packed GC column containing large particles. A downstream restriction throttled flow. The solutes were colored and were observed visually as they migrated down the column. After the separation, the fluid was bubbled through liquid solvent to collect fractions that were analyzed off-line. There was no in-line detector. This extremely crude approach is indicative of the state of the art. The results were preliminary, but were never completed by Klesper. Klesper was at the time a research assistant under Prof. Corwin, at Johns-Hopkins, in Maryland. Nevertheless, Klesper is generally considered the "father" of SFC. Klesper did not publish further on SFC until the mid-1970s.

The details on the separation of porphyrins were clarified somewhat in several later reports by Karayannis and Corwin [11], which included a number of chromatograms. Perkin Elmer had produced a "hyperpressure gas chromatograph" which replaced the previous apparatus. It was apparently extensively modified by Corwin and included a spectrophotometer as detector. The instrument does not appear to have been developed commercially as there are only a few references to its use.

Columns were typically 43 in. long (≈110 cm) with one-eighth inch ID (3.2 mm), packed with supports of Chromosorb 60–80 mesh, or 80–100 mesh. These are equivalent to 177–250 μm and 149–177 μm, respectively. These supports were coated with various stationary phases commonly used in GC. These columns probably had very low pressure drops.

The same chlorofluorocarbons were used as the mobile phases. Pressures were ≈75–200 bar. Temperatures were 140–170 °C, but at unknown densities. There appears to have been no independent means to control pressure and flow. Pressures were reported with an associated flow rate. Flow rates were between 324 mL/min at 75 bar and 756 mL/min at 180 bar. These appear to be the expanded flows measured at atmospheric pressure. The range suggests the use of a fixed restrictor to provide a pressure drop. Peaks were minutes wide.

1.1.2 Sie and Rijnders

Sie and Rijnders [12] at Shell, in Amsterdam, published a number of reports between 1966 and 1969, and they were the first to call the technique "supercritical fluid chromatography," although almost all their CO_2 data were collected under subcritical conditions. They do appear to have been the first to use a flame ionization detector (FID) in SFC which was invented the same year Golay produced his theory on capillary columns (1956). Initially, they used mostly CO_2 as mobile phase, and typically 50/70 mesh (210–297 μm) Sil-O-Cel C22 firebrick as stationary phase in a 1 m × 6 mm ID column. At that time, broken-up firebrick was a common support in packed column GC. In one report they used uncoated 120–140 mesh (105–125 μm) alumina. All these columns would have negligible pressure drops, except at high flow rates.

They immersed a cylinder of CO_2 in a large water bath and heated the water to generate a higher pressure. Today, this would be illegal in many jurisdictions for safety reasons, although some still use heating blankets, where legal. The heated reservoir acted as a pressure source. Flow was "controlled" by a manual metering valve mounted downstream of the column. Most of the pressure drop in the system probably occurred across this metering valve. All the early papers can be characterized as providing an amazing lack of detail.

Maximum pressures were ≈80 bar. Temperatures range between 30 and 40 °C. Plots of log k versus pressure, between 10 and 80 bar (40 °C), produced continuously decreasing retention that were nearly linear until they approached P_c, where the decrease accelerated. The only data point on these plots that was supercritical was at 80 bar. They generated the data as a bridge between GC and supercritical fluids. This work was really high-pressure GC, showing that CO_2 acts as a solvent (at least for low polarity solutes) even at quite low pressures (like 10 bar) and low temperatures.

At 40 °C, the density at 10 bar is 0.0177 g/cm³. At 70 bar, still below the critical pressure, the density increases more than 10-fold to 0.198 g/cm³. At 80 bar (supercritical), the density only reaches 0.281 g/cm³. These have very low densities and explain the use of light hydrocarbons as solutes. Solutes were paraffins and light hydrocarbons, which are quite volatile and easily separated by GC.

Later they switched to pentane and isopropanol (IPA) [13, 14] since these fluids were much easier to work with, being liquids at room temperature. They used the same columns in an air bath oven, but at much higher temperatures (200–250 °C). The liquid mobile phase was placed in a reservoir and pressurized with nitrogen gas at up to 50 bar. This led to some nitrogen being dissolved into the mobile phase. The densities used were ≈0.22 g/cm³, but the authors seemed to be unaware of the actual densities.

The mobile phase temperature was raised above its critical temperature in a preheater before entering the column. Injection was cumbersome because modern high-pressure rotating injection valves had not yet been invented. A series of three-way valves switched the injection port in- and off-line. After the column the mobile phase was reliquefied by passing it through a tube cooled with tap water, depressurized through a metering valve (which also nominally controlled flow), degassed, and passed through a low-pressure UV detector. Most of the solutes were polycyclic aromatic hydrocarbons (PAHs). They studied the effects of pressure and temperature on retention and efficiency. They suggested that pressure programming in SFC would be similar to temperature programming in GC.

In one early paper they showed that smaller (still huge) packings produced higher efficiency. However, they later concluded that the use of smaller particles caused large pressure drops (ΔP) along the column which they viewed as undesirable. Thus, a bias against smaller particles and large pressure drops was already articulated in 1967, even though the smallest particles used were huge by modern SFC standards.

1.1.3 Giddings

J.C. Giddings has been a towering figure in chromatographic theory since the 1950s. His book *Dynamics of Chromatography* is still a classic. He spent most of his career dealing with the theory of GC but he spent several years performing what he mostly called dense GC, before switching to field flow fractionation.

He [15] used an air-driven amplifier pump that controlled pressures up to 2,000 atmospheres, but could only make stepwise increases in pressure. Such pumps amplify by pushing a large piston connected to a much smaller piston, driving the large piston with large volumes of a low-pressure gas to obtain a much smaller volume of a much higher pressure second fluid from the smaller piston. He typically used 500 µm ID columns up to 9 m long, packed with 13 µm particles. In some cases the outlet was at atmospheric pressure. In other cases, a micrometering valve was used to raise the outlet pressure.

He recognized that retention of solutes was related to their molecular weight versus the density of the mobile phase, not its pressure, suggesting the desirability of linear density gradients.

He studied and reported what he called threshold densities for a large number of compounds. This work has often been misinterpreted. He noted that transport in the mobile phase actually starts at very low densities, but depending on detector sensitivity, the concentration may not be detected until the signal becomes high enough to exceed the detector baseline noise, whereupon the signal increases steadily. He only intended to use these densities as a relative measure of solubility. Instead, they have often been interpreted as a sudden onset of solubility that occurs when the mobile phase passes from subcritical to supercritical conditions. This has been and continues to be a major misinterpretation.

Giddings also paraphrased Hannay by saying: "One of the most interesting features of ultrahigh-pressure gas chromatography would be its convergence with classical liquid chromatography (LC). A liquid is ordinarily about 1,000 times denser than a gas; at 1,000 atmospheres, however, gas molecules crowd together with a liquid-like density. At such densities, intermolecular forces become very large, and are undoubtedly capable of extracting big molecules from the stationary phase. Thus, in effect, non-volatile compounds are made volatile." This statement was made long before Snyder [16] systematized the solvent strengths of the liquids used in high-performance liquid chromatography (HPLC) as due to a combination of the effects of dispersion, proton donor, proton acceptor, and dipole interactions in 1974.

In a very influential report, published in *Science* [17], Giddings attempted to extend the use of Hildebrand solubility parameters to polar solvents such as the alcohols, and, further, to very dense gases such as CO_2. Hildebrand and Scott [18] had warned against applying their concept to polar compounds, since it only applied to dispersive forces. Despite this warning, Giddings proposed an elutropic series, where he compared the presumed elution strength of dense fluids to normal liquids based on relative reduced densities. He calculated Hildebrand solubility parameters for the alcohols, which violated Hildebrand's warning [18]. He further placed CO_2 next to liquid IPA in elution strength. He suggested that increasing the density of CO_2 from moderate to high densities by simply increasing pressure would change its solvent characteristic from a hydrocarbon to an alcohol. If true, this would have been huge. It was NOT true but nobody knew that for decades. This was a major mistake that distorted SFC development and understanding into the 1990s, and some aspects still linger. More recent measurements, with better technology, place CO_2 similar to pentane or hexane in solvent strength, but in a different solvent family.

1.1.4 Jentoff and Gouw

Jentoff and Gouw [19], working at Chevron Research in California, introduced pressure programming in 1970. There was no effective way at the time to program density. A high-resolution LC "pulseless pump" sometimes described as a "mercury displacement pump" used high-pressure nitrogen to push immiscible mercury against the mobile phase, forcing the mobile phase out of the pump, into the column. This intervening mercury prevented dissolution of nitrogen in the mobile phase. The pump apparently was an amplifier pump, where 2,000 psi (133 bar) nitrogen could be used to generate pressures up to 5,000 psi. The fluid was then heated and passed through an injector and the column. Injection was still difficult and manual with multiple isolation valves. The pressure programmer used a high-pressure gas regulator equipped with a reducing gear train. A stepper motor rotated one gear that drove another gear mounted on the shaft of the regulator. The rate of pressure programming was controlled by changing the frequency of the driving pulses to the stepper motor. It's not clear if this was the same as a similar approach using bicycle gears with a chain used by several others, including Klesper, later in the 1970s. Caude and coworkers [20] used a motor to drive a mechanical back-pressure regulator (BPR) as late as 1991 (no pressure feedback).

Typical of the time, the columns were long (4 ft, 1.22 m), 1/4(6.4 mm) ID packed with large 100/115 mesh alumina. The mobile phase was n-pentane with 5% methanol (isocratic). The main pressure drop in the system was thought to occur in a 120 cm piece of 0.02″ ID tubing mounted downstream of the column, followed by a microregulating valve, which undoubtedly also significantly decreased pressure. Detection was performed with a low-pressure flow-through detector cell in a Cary spectrophotometer. There does not seem to have been a pressure gauge at the column outlet before the restriction. Thus, the pressure drop across the column was unknown. The only flow control was provided by the regulating valve. The flow rate and the outlet pressure would have increased as the pressure at the pump increased.

The authors produced some of the earliest separations of oligomers, in this case several polystyrenes. They also suggested that smaller particles could lead to increased retention (due to decreasing density down the column?), and thought pellicular packings, the forerunner of porous shell particles, should exhibit less resistance to mass transfer.

In later work [21], they eliminated the mercury and replaced it with a piston with spring-loaded Teflon seals (as in modern HPLC and SFC pumps), but still driven by high-pressure nitrogen. This is likely to result in a pressure drop in the pump due to friction. They also built a high-pressure flow cell capable of operation up to 4,300 psi (287 bar). They suggested that decreasing pressure/density gradients down the column would result in "disastrous consequences as far as column efficiency is concerned."

1.1.5 Novotny

Novotny et al. [22] asserted as a starting point: "uniform conditions should be carefully maintained along the entire length of the column and unpredictable changes in column pressure conditions limited." It seems reasonable that maintaining constant (density) retention is important for studying the thermodynamics of the interactions, but it obscured a bigger picture. This bias had negative consequences in terms of his conclusions.

It appears that an unspecified commercial LC (perhaps a syringe pump) was modified to perform SFC with pump flow control. This time frame corresponds to the earliest days of high-pressure LC where syringe pumps were common. The mobile phase was pentane but in a few cases 0.1% and 0.5% methanol was added. These could have been pumped reasonably accurately with a syringe pump. The mobile phase was either at room temperature (LC) or was heated to up to 220 °C. A commercial Valco injection valve was used to introduce the samples. The modifications included placing an unspecified length and ID of capillary tubing in a cooling bath after the column to reliquefy the mobile phase. A manual metering valve was placed after the capillary and before the UV detector to control outlet pressure (not flow). The UV detector operated at atmospheric pressure. Precision pressure gauges monitored the column inlet and outlet pressures. Flows were reported as below 1 mL/min measured volumetrically after the detector.

Once again, columns were long (1.5 m) with 2 and 3 mm ID, and were dry packed. Depending on the study 37–50 μm Corasil or 120–150 mesh (100–125 μm) and 80–100 mesh (149–177 μm) Porasil were used. Solutes were low molecular weight, moderately volatile compounds such as naphthalene and biphenyl. Benzene was used as a void volume marker. Values of k were calculated from the relative retention times of the solutes versus the retention time of benzene.

It was found that, although (relatively) smaller particle diameter produced better efficiency in LC, they produced worse efficiency in SFC. Further, retention significantly decreased with increasing flow rate particularly with smaller particles which was considered a serious drawback for SFC. This is probably caused by increased column head pressure, and density, at higher flows. There were no density measurements.

All early efficiency measurements must be viewed with some skepticism. The columns were typically dry packed. The packing material would be added while the column was being tapped to settle the bed. With 35–50 μm Corasil-1, the minimum height equivalent to a theoretical plate (HETP) observed in the LC mode (liquid pentane) was 2 mm (!). In SFC (supercritical pentane, density ≈0.3 g/cm³) the minimum observed HETP was 6 mm. These results were typical of all the early work by all the other authors. Measured efficiencies were very poor. Modern theory would suggest a minimum near $2d_p$ or <100 μm. Thus, all the efficiency measurements were far from optimum. Further, modern concepts relate optimum flow to the inverse of particle diameter. Following this, the optimum flow on these particles should be below 0.04 mL/min, but many times they operated at this flow rate.

1.1.6 Rogers

Nieman and Rogers [23] accepted Giddings (and Hannay's) findings that solutes had a "threshold density," where they began to migrate. A plot of such threshold densities versus molecular weight of a series of hydrocarbons was linear. Rogers also concluded that a linear density program would be superior to a linear pressure program, providing maximum resolution in a homologous series.

They developed one of the better described home-made instruments, which shows how difficult it was to perform SFC experiments, even by the mid-1970s. At this time, there was rapid progress in HPLC instrumentation, with substantial improvements in electronic and computer control.

Their first chromatograph consisted of an amplifier pump driven by 0–100 psi compressed air. An 800 step per revolution stepper motor drove an air regulator on the control side of the pump to perform pressure programming. In order to dampen the large swings in pressure generated on each stroke, five (Bourdon) pressure gauges were mounted after the pump to provide dampening. They were followed by a metering valve that dropped the pressure 500–700 bar below the pump pressure. This helped dampen pressure oscillations. After the metering valve were a series of three fine restrictors, each restrictor was followed by another pressure gauge, creating a series of pneumatic resistor/capacitor (RC) filters. This cascade decreased pressure noise from ±15 to ±2 psi. A preheater and the column were both mounted in an oven. An injection valve was mounted between the preheater and the column but outside the oven where it was isolated from the mobile phase flow and pressure by an upstream and downstream valve. The loop of the valve could be loaded at atmospheric pressure then switched online by briefly switching the two valves.

Columns were packed with 100/150 mesh, 100/120 mesh, 80/100 mesh supports with various stationary phases, in 1 m × 4.5 mm stainless steel tubes. After the column the mobile phase passed through a water-cooled tube to reliquefy the pentane, then through a high-pressure UV flow cell for detection. A micrometer metering valve, mounted after the cooling tube was used to set the initial flow and maintain the column pressure. In such a system, increasing the inlet pressure, as in a pressure program, will result in increased mass flow, increased linear velocity, and a larger pressure drop across both the column and the metering valve. Both the column inlet and outlet pressure will increase. There does not appear to be a means of measuring the column outlet pressure.

In another paper, Rogers et al. [24] attempted to reproduce the work of Jentoff and Gouw on the separation of polystyrenes using pressure programming. They used both linear and nonlinear pressure programs. The latter were used to try to mimic the density versus pressure profile of the pentane mobile phase. They actually plotted the density of the pentane versus pressure at several temperatures. The densities used were between 0.1 and 0.3 g/cm³.

Two different pumping systems were used. In both systems, modern electronic circuitry was beginning to be used to control the instrument and collect data, such as analog-to-digital converters and a digital computer, although strip-chart recorders were still used.

One system was similar to that described previously. The second system used a Varian syringe pump that was modified to act as a pressure source, and also was capable of pressure programming. A pressure gauge monitored its output. A single six-port valve acted as injector (without the isolating valves). Most of the rest was similar to the first system.

Much of the work used pure n-pentane but in some cases isopropyl or methyl alcohol was added to pentane. All mobile phases were pumped isocratically with a single pump. Flow rates were measured with a burette (with poor reproducibility). Columns still were made with 120–400 μm particles and were 1–4 m long.

In what appears to be one of the earliest attempts to control flow rate and outlet pressure independently, Simonian and Rogers [25] attempted to fix the column outlet pressure with an unusual device. They precharged a reservoir of four bottles with helium to the desired column outlet pressure, using a helium cylinder and a forward pressure regulator. The reservoir was then connected to the column outlet through a condenser tube and an in-line high-pressure detector flow cell. The reservoir was large compared to the flow rate, so the outlet pressure only increased very slowly as the mobile phase was pumped into it. The Varian 8,500 syringe pump was used as a flow source, pumping pentane.

In another paper, published in 1980 [26], they used a syringe pump to control pressure and a pair of metering valves to more accurately control flow. They acknowledge that such an arrangement resulted in an increase in flow with increased pressure. More importantly, the authors state they had not been able to separate any polystyrene oligomers using "HPLC microparticulate columns (5 and 10 μm)" with pressure programming and they returned to using much larger particles. They also used 203 μm glass beads. They came to the conclusion that larger particles yielded higher resolution, due to lower pressure drops, allowing longer columns. This may have been misinterpreted by others later.

In 1977, Klesper and Hartmann [27] stated that "one of the main obstacles in developing supercritical fluid chromatography has been the design of an apparatus which allows to keep constant or to program precisely and independently the parameters of separation i.e., pressure, temperature, flow rate, and solvent composition." They proceeded to build a complex chromatograph, with three zones of decreasing pressure control. Two solvent reservoirs were heated to increase feed pressure in order to avoid cavitation in the pumps. A double-headed membrane pump was used, which is capable of delivering two different solvents simultaneously at different flow rates, including programmed gradient elution, at up to 500 bar (flow and composition control). These pumps delivered against a high fixed back-pressure (higher than the column inlet pressure), so mass flow did not vary with column head pressure (constant compressibility). The pressure from the pumps was controlled by a metering valve with pressure feedback and a motor drive.

A similar pressure regulator control was mounted between two heat exchangers downstream of the analytical column. It could be used to fix or program column outlet pressure up until the column head pressure approached the pump delivery pressure. This was the first time both the pump flow rate and the column outlet pressure were precisely, simultaneously, and independently controlled. A third BPR was mounted downstream of the high-pressure flow cell of the UV detector. This BPR was set at a still lower pressure, compared to the column outlet pressure, and assured that the pressure in the detector cell stayed constant to avoid baseline issues due to pressure changes (refractive index changes). The column was independently heated in an oven, and temperature could be programmed.

This chromatograph was a (largely ignored) milestone in the development of packed column SFC. In some ways this design is superior to all modern instruments since it isolates the various zones from pressure perturbations in the others. However, this instrument was only appropriate for use with solvents that are liquids at room temperature and atmospheric pressure such as pentane and isopropanol. A modern instrument could be built using similar concepts but it would likely be quite expensive, and probably not necessary due to today's superior understanding of theory and physical chemistry, plus microprocessor-controlled pumps. The instrument was still used with larger particle diameter (37–75 μm) supports, in 2–6 m long columns. Most of their work was with pressure programming for separation of oligomers.

1.1.7 Capillary SFC

A dramatic change occurred in early 1981 with the publication of the first report of capillary SFC by Novotny et al. [9] in analytical chemistry. In their introduction, it was stated that "A very serious limitation of supercritical fluid chromatography with packed columns is the pressure gradient generated by the column packing...." They also referred to Jentoff and Gouw's characterization that a large decreasing density gradient down the column is the same as a negative temperature gradient in GC causing "disastrous consequences as far as column efficiency is concerned." With this background, they chose to operate a capillary column with a maximum ΔP of no more than 1 atmosphere in order to avoid any density gradients. This nevertheless allowed for very long columns.

They also referred to Giddings use of Hildebrand solubility parameters and it appears that they accepted his elutrophic series [17], placing the solvent strength of CO_2 at high pressures similar to liquid isopropanol. All this seemed to be validated by the previous 20 years results by Sie and Rijnders, Giddings, Jentoff and Gouw, Novotny, and Rogers. Small particles were thought to be bad. Pressure drops were thought to be bad. Since capillaries had minimal ΔP they avoided such problems. High-density CO_2 was thought to be as polar as an alcohol. Pressure/density programming was thought to cover a wide range of solvent strength.

In their first preliminary report they used a 200 μm × 58 m glass capillary column with a modern bonded polysiloxane stationary phase and n-pentane at constant pressure as mobile phase. They indicated that pressure programming was preferable, but not yet available. Seven PAHs were separated in ≈1 h.

They made rapid progress in the same year [28]. They demonstrated pressure programming using a syringe pump, and introduced high-pressure nitrogen through a tee after the column and before a restrictor. By changing the N_2 pressure they could vary mobile phase flow rate. However, this flow control was dropped later. They developed through the wall fluorescence detection, and still used Pyrex glass capillaries. Several high-efficiency chromatograms resembling capillary GC, but with run times of several hours, were presented. Shortly thereafter, they adopted fused silica although they claimed similar results with glass. The flexibility of the fused silica made it easier to use.

Capillary SFC was commercialized around 1984 with a syringe pump performing pressure programming, mostly with pure CO_2, and an FID, although several other detectors were developed. A fixed restrictor mounted in the base of the FID throttled flow. Most applications even with CO_2 were performed with high temperatures where the density versus pressure curves become much more linear and shallow.

It was a nearly instant success. The hardware was much simpler, and much easier to use than anything that had gone before. It was still widely thought that, at high densities, CO_2 could be as polar as IPA, and some still thought that solvation was only a matter of molecular closeness. Most of the rest of the 1980s and the first half of the 1990s was dominated by capillary SFC. There are many hundreds of chromatograms of the separation of various homologous series using capillary columns and pressure programming. This was indeed an extension of GC to much higher molecular weight compounds of limited volatility. However, there were also concerted attempts to separate relatively polar small molecules but with limited success.

Lee and Novotny received a US patent for capillary SFC, which resulted in a bit of a fight over what Lovelock meant when he suggested columns "like Desty used." In the end Lovelock's words were deemed too vague. Thus, Lee Scientific was the only company allowed to sell capillary SFC in the USA, although a number of small companies sold similar instruments, by selling their instruments for use with "micro-packed" columns.

1.2 Controversies and the birth of modern SFC

At the same time of the first report of capillary SFC, a revolutionary change in packed column SFC also occurred. A series of papers were presented by Hewlett Packard at the Pittsburg Conference in 1982, quickly followed by publications in analytical chemistry. These papers attempted to change the direction of SFC research through the presentation of substantial data that contradicted most of the then prevailing opinions. The result was a 10+ year set of controversies which eventually led to present-day SFC.

A commercial HPLC was modified slightly to pump liquefied gases [29] such as CO_2, N_2O, and NH_3 using two reciprocating pumps, as flow controllers, with the first use of chilled pump heads. These pumps had minimal compressibility compensation. A table was generated relating set flow versus actual flow for carbon dioxide at various pressures. This allowed reasonably accurate estimation of actual flows and compositions (a simple form of compressibility compensation). A commercial loop-style injection valve allowed manual injections. A UV detector cell was modified to withstand 400 bar. A mechanical BPR was used to control column outlet pressure, but was not capable of pressure programming. Both the inlet and outlet pressures were monitored with large gauges. Typical operation involved gradient elution. The whole system was controlled by a microprocessor-based chromatography workstation. Operation was the same as with HPLC. This system was simpler than Klesper's and more robust. However, it also allowed precise, independent control of flow, composition, temperature, and outlet pressure.

The real revolution was in the columns used. Gere used modern 10, 5, and 3 µm reversed-phase HPLC columns 15–25 cm long, using pure and modified fluids as the mobile phase. These columns produced substantial pressure drops. However, contrary to all previous SFC results, it was shown that plate heights decreased when the particle diameter was decreased. Further, the plate heights observed in SFC were the same as with the same column in HPLC ($h_r \approx 3$). The difference between HPLC and SFC was that the SFC optimum linear velocity was >5 times the optimum in HPLC. The SFC van Deemter curve for 3 µm particles was so flat that linear velocities up to 10 times the optimum in HPLC produced minimal increases in plate height. All this was achieved by ignoring the ΔP across the column, which had previously been considered completely unacceptable [9, 12, 21, 22, 26]. Perhaps the telling comment by Gere was "(pressure drop) ... seems to be not a problem if the column pressure drop does not occur in a region where there is a significant density differential from the column inlet to the column outlet." This comment is mirrored in current theory.

All previous work had attempted to keep the density as low as possible to maximize speed, and often used conditions very near the critical point. However, at low densities, and temperatures, density versus pressure curves can be steep, generating large gradients in density and linear velocity if there is even a relatively modest pressure drop across the column. For example, at 40 °C, the density of CO_2 changes from ≈ 0.28 g/cm³ at 80 bar to 0.62 g/cm³ at 100 bar. This represents ≈ 0.017 g/cm³/bar.

Gere simply worked higher on the pressure versus density curve. For example, with the column outlet pressure set to 100 bar, the column inlet pressure might be 150 bar. Density would change from ≈ 0.62 to 0.78 g/cm³, or average ≈ 0.0032 g/cm³/bar. This represents a density gradient less than one-fifth the gradient between 80 and 100 bar.

Gere's work resulted in a kit commercially sold by Hewlett Packard to convert HP1082 and 1084 HPLCs into SFCs. This was the first commercially available SFC (it was sold before the Lee Scientific capillary SFC). Less than 50 were sold before the SFC incompatible 1090 replaced the 1084, but most of them were still in service after 10 years. Since this work, virtually all subsequent packed column SFC reports have used columns similar to those used in HPLC, most typically 10–25 cm long packed with sub-2–5 µm particles.

This chapter marked the change in perception that SFC is an extension of GC to lower polarity, low volatility, larger molecules, to an extension of HPLC with an alternate selectivity for more polar molecules with enhanced speed and lower pressure drops, compared to HPLC. Some view SFC as a bridge between GC and HPLC, sometimes calling it "unified chromatography" [30].

It should be noted that most of Gere's work was performed with pure CO_2 to avoid any issues or controversy regarding binary fluids. In a few instances, binary fluids were used. However, these were not generated by pumping a normal liquid with one pump and CO_2 in the other. A small weight of modifier was added to a 1 L high-pressure reservoir, which then had a weight of CO_2 added. The mixture was equilibrated and pumped as one of the fluids. The other pumping channel sometimes delivered pure CO_2 in order for the final fluid to be accurate with less than 1% modifier.

As an aide, after this work, several gas vendors started selling large-volume, pre-mixed cylinders of CO_2 with various concentrations of modifier, since it was thought that the composition was fixed, eliminating uncertainty in composition, and requiring only a single pump, particularly for supercritical fluid extraction (SFE). However, it was later shown [31] that the headspace and the liquid phase had different concentrations of modifier, resulting in a gradual change in composition as the cylinder was used up. Such an approach was abandoned.

Lauer et al. [32], also from Hewlett Packard, measured binary diffusion coefficients (D_{12}) for a number of relatively nonpolar solutes in pure CO_2 at various densities. The temperature range used bracketed the critical temperature. Plots of ln D_{12} versus $1/T$ produced straight lines, regardless of whether the density was kept constant, by varying the pressure, or when pressure was kept constant allowing the density to vary. Since some of these measurements were subcritical and some were supercritical, the fact that there was no deflection at T_c indicates that most of the work in SFC, up until this time, was performed under a mistaken impression that diffusion in the fluid changed significantly when passing from subcritical to supercritical conditions. There are still many users who believe something important happens when the name changes from subcritical to supercritical. Not a single international symposium passes without someone asserting that such happens although it clearly does NOT.

These authors also made van't Hoff plots of log retention (k) versus $1/T$ for a number of low polarity solutes, plus caffeine, in both CO_2 and N_2O, with temperatures bracketing T_c, at a constant density of 0.8 g/cm³. The plots were all linear. These measurements together indicate there is no change in the nature of either retention or diffusivity when crossing from subcritical (liquid) to supercritical conditions, which was counter to the suppositions in all previous work.

Most of the early works accepted that there was a "threshold density" below which a solute would not migrate. However, an additional restraint was often imposed. It was thought that the temperature must be above T_c for migration to commence above P_c. Many authors seemed to convolute solubility with diffusivity and retention. Solubility is required in both HPLC and SFC (at least for nonvolatile solutes). However, once a compound is migrating it appears that retention and diffusion are only modest functions of density.

Weaver and Bente [33] at Hewlett Packard reported brief solvatochromic dye measurements that showed pure CO_2 was similar to hexane in solvent strength but apparently did not publish the results in a journal. They [34] also recalculated Hildebrand solubility parameters for CO_2 which confirmed this low value for its solvent strength. Both sets of data were presented at the fall meeting of the American Chemical Society, Physical Chemistry Division in 1983.

Frank [35], also at Hewlett Packard, studied the effect of modifier on solvent strength and selectivity. She used Snyder's selectivity triangle and P' solvent strength scale to characterize the differences in solvents as proton donors, proton acceptors, and dipoles.

In 1985 Jasco introduced a combined SFE–SFC instrument with reciprocating pumps having a chilled pump head, and a programmable electronic BPR. This was the first low internal volume electronic BPR designed specifically for SFE and SFC. By this time people were using one pump to deliver pure fluids, usually CO_2, and the other to pump modifier. This was the only commercial packed column SFC until 1992, since the HP 1084 was withdrawn in 1984. Jasco is the only company to continuously offer SFCs for the last 32 years.

1.2.1 Packed versus capillary

The almost simultaneous invention of capillary SFC and the rebirth of packed column SFC using small particles inevitably resulted in a serious controversy about which was "better." The combined work of Gere, Lauer, Bente, Weaver, and Frank pointed toward a different path, compared to everything that had happened before. Capillary SFC seemed to validate everything that had gone before. There developed a great deal of hostility, particularly against the few remaining proponents of packed columns. This got so bad that Dai Games, a very prominent LC-mass spectrometry (MS) and SFC-MS researcher, had an oral presentation about packed column SFC-MS rejected by Pittcon! Pittcon is the place where graduate students make their first oral presentations, for a person of his prestige to be rejected was ridiculous.

Three notable groups continued to publish packed column SFC from the mid- to late 1980s. Caude and coworkers were very active in the separation of many relatively polar compounds including opioids. They also published the first chiral separations [36] by SFC. Of course chiral separations have become the mainstay of SFC at both the analytical and semiprep scales. Dai Games published a number of biochemical SFC-MS applications such as conjugated bile acids, mycotoxins, veterinary drugs, and sulfonamides. The group at Ciba Geigy in Switzerland was probably the largest industrial user, with a huge group developing methods, and to a lesser extent instrumentation. Their work eventually led to the Gilson SFC, introduced in 1992. However, most other workers switched to capillary SFC.

Even into the late 1980s most workers still believed that CO_2 was as polar as IPA due largely to the enormous stature of Giddings. With such high solvent strength it was reasoned that capillary SFC was capable of eluting quite polar solutes, leaving little room for packed column usage, which was much more complicated and expensive and at the time had problems with column inertness. Many also believed that the changes in retention often observed when alcohols were added to the mobile phase were only the result of increasing the density of the fluid, despite the fact that no density measurements of such fluids existed.

Subsequently, Berger [37, 38] measured the density of CO_2/MeOH mixtures and proceeded to measure k as a function of modifier concentration while keeping the average density in the column constant. The results clearly showed that increasing the modifier concentration dramatically increased solvent strength and decreased retention.

Berger and coworkers [39, 40] also used solvatochromic dyes to measure the solvent strength of CO_2/modifier mixtures and the effects of additives, as a function of modifier concentration. CO_2 was shown to be as nonpolar as pentane. This was more or less a wake-up call for most who had believed that CO_2 was much more polar than it actually was. It was also found that the first small additions of polar modifiers, such as methanol, caused dramatic increases in solvent strength.

Another issue dogged packed columns through the 1980s. It had often been true that many acids and bases either tailed badly, or did not elute even with high concentrations of modifier. This was usually blamed on active sites on silica supports. A major step forward was the introduction of polar additives to the modifier. In the first such reports [41, 42], peak shapes of phenylthiohydantoin (PTH)–amino acids were improved dramatically. Since then many more studies on the use of additives have been made. The use of additives has expanded the use of SFC to much more polar solutes, including polyacids and primary amines, and even some peptides. More recently, there have been concerted efforts to avoid additives, particularly when performing SFC-MS. This has resulted in a number of new stationary phases, which in some cases do not require additives.

1.2.2 Schoenmakers

In the mid-1980s, Schoenmakers [43–45] published several papers, and a book chapter, on the effect of pressure drops on efficiency across packed columns. He modified an LC pump to operate at constant column inlet pressure, and used a Tescom mechanical BPR to control outlet pressure. However, he used inlet pressure as low as 102 bar, while controlling the pressure drop to greater than 20 bar (exactly the range with the steepest density gradients). In one case the inlet pressure was 122 bar, with the outlet pressure adjusted to get a ΔP of up to 45 bar. Thus, the outlet pressure must be as low as ≈77 bar. At 40 °C, the density of CO_2 at 122 bar is >0.723 g/cm³. At 77 bar, the density is ≈0.248 g/cm³. Under such conditions the density gradient across the column was excessive, and linear velocity increased by a factor of 2.9 across the column. He observed severe efficiency losses with ΔP >20 bar. Although these were extreme conditions, he generalized his finding to include all pressure drops at all densities, despite Gere's comment [29] about the need to minimize density gradients across the column. He concluded that packed column SFC was limited to ΔP < 20 bar. This would limit SFC with 5 µm particles to <20,000 plates, and much less with smaller particles. Poe and Martire [46] and Mourier et al. [47] had similar theories.

Shoenmakers results seemed to fit well with most of the early works by Sie and Rijnders, Jentoff and Gouw, Novotny, and Rogers. All of them tended to work at as low a density as possible to maximize the speed advantage of the fluids in terms of highest diffusion coefficients. Those earlier workers had all used very large particle supports with minimal pressure drops. When smaller particles were used they tended to see degraded performance. However, in such regions, even relatively small ΔP's tended to create large density gradients, and large changes in linear velocity across the column. Tarafder and Guiochon [48] have explained the effect.

In response to Schoenmakers findings that any ΔP >20 bar caused severe efficiency losses, Berger and Wilson [49] connected up to eleven 20 cm long columns packed with 5 µm particles in series creating a 2.2 m long column that produced a ΔP of ≈160 bar at 10% methanol at 2 mL/min, 40 °C, outlet pressure 130 bar. This stack of columns produced over 220,000 plates (h_r = 1.98), thoroughly debunking the idea that ΔP >20 bar always caused excessive efficiency losses, and that packed column SFC was limited to <20,000 plates. It is nonetheless true that under conditions where density changes dramatically with small changes in pressure there will be serious efficiency losses. One merely needs to avoid such conditions.

This was the last major theoretical hurdle preventing further development of packed column SFC into what we see today.

1.2.3 Second generation

In 1992, Hewlett Packard (HP, now Agilent), and Gilson introduce second-generation packed column SFC instruments. The HP instrument had dynamic compressibility compensation for accurate flow and composition, independent flow, composition, temperature, and outlet pressure control. The user could program flow (0.1–5 mL/min), composition (0–100% modifier), temperature (–50° to 150 °C), outlet pressure (40–400 bar), or density (pure CO_2 only). Diode array and FIDs were the norm but more exotic GC detectors such as the nitrogen-phosphorous detector (NPD) and electron capture detector (ECD) were available. An autosampler allowed 100 samples to be run automated. A PC-based ChemStation controlled the instrument and collected data. The instrument could be configured to also perform capillary SFC, but was primarily a packed column instrument. It tended to be used in the separation of polar solutes.

About this time, the limitations of pure CO_2 and capillary columns were being recognized, particularly when compared to packed columns with the more polar solutes. It became clear to most users that pure CO_2 was far less polar than previously thought. Further it was far slower than small particle packed columns, and had reliability issues associated with the need for tiny fixed restrictors in the FID that often plugged or changed resistance/flow over time. By the end of the 1990s capillary SFC had practically disappeared. This is a shame since it was clearly superior for a number of applications particularly for extending GC to higher molecular weight, lower volatility solutes, especially oligomeric series. Many such oligomeric series now run on HPLC were better done with capillary SFC. The reality was that capillary SFC was indeed useful for extending GC to larger, less volatile solutes. Packed column SFC has been shown to be most useful as providing a faster alternate selectivity to HPLC for moderate to high polarity solutes.

Jasco has continued to evolve its SFE and SFC equipment. This author bought the SFC business from Hewlett Packard at the end of 1995 and established Berger Instruments (BI). BI developed the first commercial SFC-MS, the first chiral method development system, a series of the first successful semiprep SFCs, and gas delivery systems. BI was sold to Mettler-Toledo AutoChem in 2001 and was later sold to Thar. Waters subsequently purchased the SFC part of Thar.

1.2.4 Third generation

Berger reassembled much of the BI R&D team and formed Aurora SFC Systems in 2007 and produced a conversion module for Agilent 1100s and 1200s, converting the HPLCs into SFCs. Agilent bought Aurora in 2012 and continues to improve the product. Waters introduced what they call UPC² in 2009. Recently, Shimadzu introduced a combined SFE/SFC. Jasco continues to improve their product.

Since the early 1990s most improvements in instruments have been largely iterative rather than dramatic. Today's pumps and BPRs are much better than in the past but there is still significant room for improvement. There are now four major and a few smaller instrument companies selling SFC equipment. A few features have disappeared, but generally performance has improved. Pressure (or density) programming was a major feature in the 1990s for the analysis of homologous series. This almost completely disappeared in the 2000s. However, both the Waters and Agilent SFCs have reintroduced pressure programming, although few applications of the old style have emerged. However, large composition gradients from nearly pure CO_2 to nearly pure liquid modifiers are emerging where reversed pressure programming appears to be useful. The FID was also very important, particularly with pressure programming involving pure CO_2.

There continues to be a preoccupation with the use of low-density CO_2 to maximize the speed of analysis. However, such conditions are only useful for the separation of compounds easily separated by GC and tend to result in significant density gradients. SFC continues to be most useful for somewhat polar molecules such as small drug-like molecules and should be the technique of choice for chiral separations.

What lies in the future is difficult to say. It appears that SFC has found a place in separation science and is likely to keep expanding and being improved. The idea of "Green Chemistry" is finally widely accepted as real and important. There are many indications that the dramatic success SFC has seen in chiral separations is beginning to spill over into achiral separations.

Probably the biggest issue today is the high cost of entry. Most universities, particularly in the USA, do not teach SFC or only do so without hands-on experience due to the cost of the equipment. Along the same line, although there are relatively inexpensive, less capable GC and HPLC available, there are no "lower end" SFCs. This is undoubtedly due to the relatively smaller volumes the vendors sell. This impacts their level of investment in R&D and training, and in the cost of goods incorporated into their products. It is hoped that as the market expands, list prices will fall.

References

[1] Cagniard de la Tour, C, Exposé de quelques résultats obtenus par l'action combinée de la chaleur et de la compression sur certains liquides, tels que l'eau, l'alcool, l'éther sulfurique et l'essence de pétrole rectifiée, Annales de chimie et de physique, 1822, 21, 127–132.

[2] Andrews, T, The Bakerian lecture: On the continuity of the gaseous and liquid states of matter, Philo. Trans. Roy. Soc. (London), 1869, 159, 575–590.

[3] Hannay, J., Hogarth, J. On the solubility of solids in gases, Proc. Roy. Soc. (London) 1879, 30, 178–188.

[4] Krukonis, V, European Pharmaceutical Contractor (EPC) May 1998 (may no longer be available except through http://www.phasex4scf.com/supercritical-fluids-applications).

[5] Hannay, JB, On the solubility of solids in gases, Proc. Roy. Soc., London, 1880, 30, 484–489.

[6] Buchner, EG, Die beschrankte Mischbarkeit von Flussigkeiten das System Diphenylamin und Kohlensaure, Z. Phys. Chem., 1906, 56, 257.

[7] Golay, M. J. E., Vapor phase chromatography and telegrapher's equation. Anal. Chem., 1957, 29, 928–932.

[8] Desty, DH, Capillary columns: Trials, tribulations and triumphs, in J.C. Giddings, R.A. Keller, eds, Advances in Chromatography, New York, Marcel Dekker, Inc., 1965, p. 199.

[9] Novotny, M, Springston, SR, Peaden, PA, Fjeldsted, JC, Lee, ML, Capillary supercritical fluid chromatography, J. Chromatogr., 1981, 61, 17–28.

[10] *Klesper,* E, Corwin, AH, Turner, DA, High pressure gas chromatography above critical temperatures, *J. Org. Chem.,* 1962, *27,* 700–701.

[11] Karayannis, NM, Corwin, AH, Hyperpressure gas chromatography. III. Gas chromatography of porphyrins and metalloporphyrins, Anal. Biochem., 1968, 26, 34–50.

[12] Sie, STE, Rijnders, GWA, High pressure gas chromatography and chromatography with supercritical fluids. 1. The effect of pressure on partition coefficients in gas-liquid chromatography with carbon dioxide as the carrier gas, Sep. Sci., 1966, 1, 459–490.

[13] Sie, STE, Rijnders, GWA, High pressure gas chromatography and chromatography with supercritical fluids. III. Fluid-Liquid Chromatography, Sep. Sci., 1967, 2, 729–753.

[14] Sie, STE, Rijnders, GWA. High pressure gas chromatography and chromatography with supercritical fluids. IV. Fluid-Solid Chromatography, Sep. Sci., 1967, 2, 755–777.

[15] Giddings, JC, A critical evaluation of the theory of gas chromatography, in Goldup, A. ed., Gas Chromatography 1964, Elsevier Publishing Co., 1965, pp. 3–23.

[16] Snyder, L., Classification of the solvent properties of common liquids, J. Chromatogr., 1974, 92, 223–230.

[17] Giddings, JC, Myers, MN, McLaren, L, Keller, RA, High pressure gas chromatography of nonvolatile species, Science, 1968, 162, 67–73.

[18] Hildebrand, JH, Scott, R.L. Solubility of Nonelectrolytes, Reinhold, New York, 1950.

[19] Jentoff, RE, Gouw, TH, Pressure programmed supercritical fluid chromatography of wide molecular weight range mixtures, J. Chromatogr. Sci., 1970, 8, 138–142.

[20] Villermet, A, Thiébaut, D, Caude, M, Rosset, R, Packed column supercritical fluid chromatography with carbon dioxide-polar modifiers. Influence of carbon dioxide density on retention, J. Chromatogr., 1991, 557, 85–97.

[21] Jentoft, RE, Gouw, TH, Apparatus for supercritical fluid chromatography with carbon dioxide as the mobile phase, Anal. Chem. 1972, 44, 681–686.

[22] Novotny, M, Bertch, W, Zlatkis, A, Temperature and pressure effects in supercritical fluid chromatography, J. Chromatogr., 1971, 61, 17–28.

[23] Nieman, JA, Rogers, LB, Supercritical fluid chromatography applied to the characterization of a siloxane-based gas chromatographic stationary phase, Separation Science, 1975, 10, 517–545.

[24] Conaway, JE, Graham, JA, Rogers, LB, Effects of pressure, temperature, an adsorbent surfaces and mobile phase composition on the supercritical fluid chromatographic fractionation of monodisperse polystyrenes, J. Chromatogr. Sci., 1978, 16, 102–110.

[25] Simonian, BP, Rogers, LB, Unusual gas chromatographic behaviors of naphthalene, pyrene, and phenanthrene using pressurized n-pentane as a carrier gas, J. Chromatogr. Sci., 16 (1978) 49–60.

[26] Graham, JA, Rogers, LB, Effect of column length, particle size, flow rate, and pressure programming rate on resolution in pressure programmed supercritical fluid chromatography, J. Chromatogr. Sci., 1980, 18, 75–84.

[27] E. Klesper, E, Hartmann, W, Apparatus and separations in supercritical fluid chromatography, European Polymer Journal, 1977, 14, 77–88.

[28] Peaden, PA, Feldsted, JC, Lee, ML, Springston, SR, Novotny, M, Instrumental aspects of capillary supercritical fluid chromatography, Anal. Chem., 1982, 54, 1090–1093.

[29] Gere, DR, Board, R, McManigill, D, Supercritical fluid chromatography with small particle diameter packed columns, Anal. Chem., 1982, 54, 736–740.

[30] Parcher, JF, Chester, TL, eds, Unified Chromatography, ACS Symposium Series, Vol. 748, American Chemical Society, 2000.

[31] Schweighardt, FK, Mathias, PM, Impact of phase equilibria on the behavior of cylinder-stored CO_2-modifier mixtures used as supercritical fluids, J. Chromatogr. Sci. 1993, 31, 207–211.

[32] Lauer, HH, McManigill, D, Board, RD, Mobile-phase transport properties of liquefied gases in near-critical and supercritical fluid chromatography, Anal. Chem., 1983, 55, 1370–1375.

[33] Weaver, HE, Bente, PF, Solvatochromic effects of mesityl-oxide and pyrazine in fluid carbon dioxide, Fall meeting of the American Chemical Society, Physical Chemistry Division in 1983. Paper 23.

[34] Bente, PF, Weaver, HE, Hildebrand, JH, Solubility parameter for liquids and supercritical carbon dioxide, fall meeting of the American Chemical Society, Physical Chemistry Division in 1983. Paper 24.

[35] Frank, LG, Column efficiencies and mobile phase solvent power in carbon dioxide based supercritical fluid chromatography, in Ultrahigh Resolution Chromatography, Chapter 11. ACS Symposium Series 250 Washington, D.C., 1984.

[36] Mourier, P, Sassiat, P, Caude, M, Rosset, R, Retention and selectivity in carbon dioxide with various stationary phases, J. Chromatogr., 1986, 353, 61–15.

[37] Berger, TA, Density of methanol-carbon dioxide mixtures at three temperatures, J. High Resolut. Chromatogr., 1991, 14, 312–316.

[38] Berger, TA, Deye, JF, Composition and density effects using methanol/carbon dioxide in packed column SFC, Anal. Chem., 1990, 62, 1181.

[39] Deye, JF, Berger, TA, Anderson, AG, Nile Red as a solvatochromic dye for measuring solvent strength in normal liquids and mixtures of normal liquids with supercritical and near critical fluids, Anal. Chem., 1990, 62, 615.

[40] Berger, TA, Deye, JF, Use of solvatochromic dyes to correlate mobile phase solvent strength to chromatographic retention in SFC, in Bright, FV, McNally, MEP, eds., Supercritical Fluid Technology, ACS Symposium Series 488, American Chemical Society, Washington, D.C., 1992.

[41] Ashraf-Korassani, M, Fessahaie, MG, Taylor, LT, Berger, TA, Deye, JF, Rapid and efficient separation of PTH-amino acids employing supercritical CO_2 and an ion pairing agent, J. High Resolut. Chromatogr., 1988, 11, 352.

[42] Berger, TA, Deye, JF. Ashraf-Korassani M, Taylor, LT, Gradient separation of PTH-amino acids employing supercritical CO_2 and modifiers, J. Chromatogr. Sci., 1989, 27, 105–110.

[43] Janssen, HG, Schijders, HMJ, Rijks, JA, Cramers, CA, Schoenmakers, PJ, The effects of the column pressure drop on retention and efficiency in packed and open tubular supercritical fluid chromatography, J. High Resolut. Chromatogr., 1991, 14, 438–445.445.

[44] Schoenmakers, PJ, Open columns or packed column for supercritical fluid chromatography- A comparison, in Smith, RM ed., Supercritical Fluid Chromatography, Royal Society of Chemistry, London, 1988, Chapter 4.

[45] Schoenmakers, PJ, Uunk, LGM, Effects of pressure drop in packed column Supercritical Fluid Chromatography, Chromatographia, 1987, 24, 51–57.

[46] Poe, DP, Martire, DE, Plate height theory for compressible mobile phase fluids and its application to gas, liquid, and supercritical fluid chromatography, J. Chromatogr., 1990, 517, 3–29.

[47] Mourier, PA, Caude, MH, Rosset, RH, The dependence of reduced plate height on reduced velocity in carbon dioxide supercritical fluid chromatography with packed columns, Chromatographia 1987, 23, 21–25.

[48] Tarafder, A, Guiochon, G, Unexpected retention behavior of supercritical fluid chromatography at the low density near critical region of carbon dioxide, J. Chromatogr. A, 2012, 1229, 249–259.

[49] Berger, TA, Wilson, WH, Packed column supercritical fluid chromatography with 220,000 plates, Anal. Chem., 1993, 65, 1451–1455.

.

Eric R. Francotte

2 Achiral preparative supercritical fluid chromatography

Abstract: Packed column supercritical fluid chromatography (SFC) has been around for more than 40 years but the adoption of this technology as a general and routine method for chiral separations and achiral purifications is much more recent. SFC is known for its spectacular development over the past decade and this revival of the technique has been mainly driven by its preparative usefulness. The green features, the speed, and the low running costs of SFC have greatly facilitated its adoption for preparative purposes, as these factors are essential at the preparative scale. With the introduction of more reliable instruments, the utilization of preparative SFC for chiral separations has rapidly expanded and it is now the method of choice for this application. Preparative packed column SFC is also now rapidly and increasingly embraced by many research laboratories for preparative achiral purifications, and in particular in the drug discovery environment and other life science sectors. Indeed, at this stage of drug development, it perfectly fits the requirements, that is, rapid processing of many samples in relatively small amounts. The recent introduction of preparative SFC instruments possessing mass-triggered fractionation capabilities has markedly contributed to the extraordinary acceleration of the expansion of the technique for achiral purifications. The numerous advantages of packed SFC such as high diffusivity, low pressure drop, short equilibration times, reduced solvent consumption and costs, less safety concerns with respect to flammability and toxicity, fast solvent removal, and reduced impact on the environment have certainly also been decisive for this major conversion from HPLC to SFC in the field of preparative purifications. SFC has now become the first approach in drug discovery for the fast purification of thousands of samples produced by high- and medium-throughput synthesis. This chapter focuses on the current status of preparative packed column SFC for achiral applications.

Keywords: preparative chromatography, supercritical fluid chromatography, achiral purification, stationary phase, SFC instruments

2.1 Introduction

Packed column supercritical chromatography has been around for more than 5 decades. The first application was reported by Klesper et al. who preparatively separated a few milligrams of nickel etioporphyrin II and nickel mesoporphyrin dimethylester mixtures using chlorofluoromethanes as supercritical fluids [1].

Eric R. Francotte, FrancotteConsulting, Switzerland

https://doi.org/10.1515/9783110500776-002

Previously, SFC was called high-pressure gas chromatography above critical temperatures [1–3], as it was not clear whether supercritical fluid should be classified as a gas or liquid chromatography process. The question is no longer debated and it is generally understood that the supercritical fluid state is a particular state (neither a gas nor a liquid). The term supercritical fluid chromatography (SFC) is not restricted to CO_2, but because almost all SFC applications use CO_2 as the supercritical fluid, it is generally implicitly assumed that SFC means SFC with CO_2. Numerous substances exhibit supercritical properties at a certain temperature and pressure [4], but carbon dioxide is the preferred element for chromatographic purposes because of number of following practical reasons: (1) critical temperature and pressure of CO_2 are compatible with common chromatographic instrumentations, (2) CO_2 shows a very low reactivity and therefore is nondestructive for the molecules to be analyzed or purified, (3) CO_2 has a low toxicity, (4) CO_2 is nonflammable and (5) is abundant, and (6) CO_2 is inexpensive.

2.1.1 Evolution of preparative SFC

Most of the earlier SFC investigations focused on analytical capillary SFC, and the first real preparative applications were patented by Perrut in 1982 using pure CO_2 as the mobile phase [5]. In the next 20 years, new reports on preparative applications of SFC for achiral purifications were published but the occurrence was relatively low. A number of semi-preparative and preparative SFC instruments were introduced by different companies between 1990 and 1999 for achiral and chiral separations. Most instruments were essentially derived from the analytical instruments that were adapted for preparative purpose and equipped with phase separators [6]. In 1991, T. Berger demonstrated the applicability of SFC to polar compounds in a series of papers describing the utilization of polar modifier and additives [7–9]. This was an important observation as it changed the perception that SFC is only applicable to nonpolar molecules and stimulated the extension of SFC to a much larger field of applications.

Although all initial earlier preparative SFC applications on packed columns were related to the purification of achiral compounds, the adoption of the technology has progressed much more rapidly in the field of chiral separations from around the beginning of the new millennium. Several of the following reasons can be imputed to this development: (1) the introduction of new and more reliable preparative SFC instruments; (2) the easy transfer from normal phase chromatography mode to SFC, which is also a normal phase chromatography mode; and (3) the fact that the same chiral columns can be used either for LC or SFC. In the meantime, chiral SFC has become a routine and the preferred method for preparative chiral separations [10–12].

The situation was different for achiral purifications as reversed-phase HPLC had established since the late 1980s as a powerful and quasi-universal technique for this purpose. However, things radically changed about 10 years ago, when dedicated preparative SFC instruments comprising mass-directed fractionation capabilities were introduced on the market. This instrumental progress coincided with the worldwide acetonitrile shortage [13–14], which forced the chromatographers to initiate plans to rapidly elaborate new strategies and look at alternative technical options to replace the popular reversed-phase HPLC. SFC could profit from this opportunity and this constellation has revolutionized the area of preparative achiral purifications. Over the last 7 years, SFC has extremely expanded in this area and has imposed as a valuable complementary and competitive option to purify *hundreds of thousands* of achiral component mixtures, particularly in medicinal chemistry.

The evolution of the development of analytical and preparative SFC instrumentation has recently been reviewed in several papers and will not be covered in this chapter. Several review articles are available on the subject [15–17]. It must be emphasized that technical problems associated with the formation of aerosol and low recovery were very discouraging at the beginning. Besides this major issue, the lack of robustness, high instrument costs, and expensive CO_2 delivery infrastructure were other major concerns explaining the slowness that has characterized the adoption of the technology in the pharmaceutical environment. Moreover, the consciousness for environmental considerations was not as developed at that time.

2.1.2 Features and advantages of preparative SFC

Considering that almost all SFC applications are performed in the presence of organic modifiers, there is still a controversy regarding the term utilized when performing SFC separations. As soon as organic modifiers such as methanol are added to supercritical CO_2, the critical point (temperature and pressure) are shifted [18–19]. For example, with 30% methanol, the critical temperature is about 86 °C and the critical pressure is about 142 bars; these values are far from the conditions that are typically applied in SFC (Table 2.1). Therefore, the term Subcritical Fluid Chromatography that was proposed by Caude and coworkers [20] is often used today. Even though the CO_2

Table 2.1: Critical temperature and pressure of carbon dioxide/methanol mixtures (adapted from [18] and reference cited therein).

CO_2 Volume Ratio	Methanol Volume Ratio	Methanol Mole Fraction	Critical Temperature (°C)	Critical Pressure (MPa)
100	0	0	31.0	7.38
90	10	0.07	51.1	10.46
80	20	0.12	64.9	12.22
70	30	0.20	85.9	14.37
50	50	0.36	124.4	16.54

chromatographic fluid has some normal phase character in this state, most of advantages of SFC (Table 2.1) remain applicable.

The beneficial characteristics of SFC such as high diffusivity and low viscosity are now fully exploited in analytical SFC using modern instruments that show similar or sometimes higher performances than those achieved by Ultra Performance Liquid Chromatography (UPLC) [21–22]. For preparative applications, SFC shows additional advantages such as huge reduction of solvent consumption and cost, faster and easier recovery of the purified substances, and reduced environmental and safety concerns compared to HPLC (Table 2.2). These benefits have greatly contributed to the re-emergence of SFC that has clearly been driven by its preparative utilization and in particular by the application to enantioselective separations. As a follow-up of this successful implementation for chiral separations, packed column SFC is currently recognized for rapid development in many industrial research organizations for achiral analysis and purifications as well. Lower energy costs, reduced waste disposal costs, and higher sample stability are additional attractive features of SFC for preparative applications. The benefits of the preparative SFC technology are summarized in Table 2.2.

Table 2.2: Preparative SFC purification technology: Main benefits.

Speed	– SFC can be operated faster than HPLC (2–4 times) due to lower viscosity
	– Column conditioning/equilibration is much faster
	– Compound recovery is much more easier and more rapid
Safety	– CO_2 (main component of the mobile phase) is nonflammable
	– CO_2 is much less toxic than most organic solvents
Costs	– Reduced solvent costs (by ~ 60%–80%)
	– Less energy costs (fraction evaporation)
	– Reduced waste disposal costs
Sample stability/toxicity	– No acidic additive needed in the mobile phase and consequently, no salt formation with basic compounds and less risk of decomposition of acid/basic sensitive chemical groups.
Efficiency	– Less pressure drop (small particle size and/or longer columns can be used)
Impact on environment	– Reduced organic solvent emission
	– Less waste disposal

Because of low viscosity of supercritical CO_2 (near to gas state), it is possible to work at very high flow rates, which cannot be reached by HPLC. As a consequence, the run times are significantly shorter and the purifications/separations are on average 3 to 4 times faster. The easy recovery process of the purified compounds is another key advantage in preparative SFC. Indeed, as CO_2 is removed almost instantaneously, the organic phase that has to be evaporated is much smaller than in an HPLC process.

Because of low viscosity of supercritical CO_2, it is possible to work with packing materials having small particle size or with longer (tandem) columns, both leading to a significant increase of efficiency. In SFC, it is usual to use 5 mm particle size for preparative applications. This higher efficiency permits to achieve good separations even when the separation presents only small selectivity values.

Safety is also a positive feature of SFC. CO_2 which constitutes the main component of the mobile phase in SFC is nonflammable and much less toxic than most organic solvents used in HPLC. Alkanes (heptane and hexane) in normal phase HPLC are neurotoxic, and acetonitrile, which is the most usual organic component in reversed-phase HPLC, is highly toxic by inhalation and by contact (skin penetration). Both factors, that is, flammability and toxicity, are particularly important in the preparative field where operators are dealing with large amounts of solvents.

In preparative HPLC, acidic or basic additives are often used to suppress undesirable peak tailing or distortion effects because of unfavorable interactions with silanol groups on the surface of the chromatographic stationary phase. As a consequence, basic compounds that are most common in the pharmaceutical field are isolated as salt (most frequently TFA). Trifluoroacetic Acid (TFA) is known to be toxic for biological cells and must be removed before testing, necessitating an additional step of purification. Usually, SFC does not require to utilization of additives and this feature considerably facilitates and accelerates the isolation of the target compound(s). Moreover, this strategy permits to avoid the possible decomposition of acid or base sensitive compounds.

Costs are of course a critical factor in preparative chromatography. Replacing organic solvents in normal phase chromatography by CO_2 permits to save 70% to 90% of the solvent costs. It is a little bit less compared to reversed-phase HPLC as water is also a cheap solvent. However, removing water to isolate the purified substances is an energetically costly process. According to our estimation [23], it needs 7 times more energy compared to the energy needed to evaporate small amounts of modifier in SFC. Additionally, waste disposal that is non-negligible cost factor is considerably reduced in SFC compared to HPLC.

SFC is often considered as a green technology as it has a positive impact on our environment. While this aspect was often neglected in the past, it has become a major concern in almost all industrial and private sectors today. In this respect, SFC clearly contributes in reducing solvent usage and organic solvent emission. Moreover, reduction of waste disposal also indirectly contributes to decrease the production of more carbon dioxide in the environment as organic solvent waste is usually burnt after usage in other chromatographic processes. In this context, it is worthwhile to stress that SFC does not produce CO_2; it just makes use of existing CO_2, mostly obtained as a by-product of large industrial processes.

2.2 Instrumentation

Preparative achiral purifications using packed Supercritical Fluid Chromatography (pSFC) were already around before 1990, but the number of applications remained limited and it was almost totally ignored by the pharmaceutical and agrochemical industry. Flash chromatography and reversed-phase HPLC had established as the gold standards.

Between 1990 and 1999, several companies (Prochrom, Gilson, Jasco, Berger, and Thar) introduced preparative SFC instruments to the market with more or less success. Most of these instruments were rather designed for small-scale separations, capable of delivering flow rates ranging between 25 and 50 mL/min. The main application field was chiral separations, although the instruments were also adapted to achiral separations. Achiral purifications were mostly limited to the separations of "simple" mixtures containing only a few components (2–3) and usually when reversed-phase mode failed.

For purification of more complex mixtures such as combinatorial chemistry libraries, the available SFC instruments were not adequate because none of them offered MS-directed fractionation. This selective capability is essential for this type of application dealing often with low purity samples containing many components. This shortcoming motivated a few groups develop their own in-house semi-preparative SFC–MS [24–26].

It took several years until this option became available on a commercial instrument when the Thar SFC-MS Prep-30 system was introduced in 2008 [27], solving a long-existing challenge and filling the gap. This new possibility has revolutionized the area of preparative achiral SFC purifications, which is now quickly expanding, especially in the pharmaceutical industry that has rapidly recognized the value of the technology in drug discovery producing hundreds of thousands of new molecules every year.

2.2.1 CO_2 delivery and recycling

The CO_2 delivery unit is an integral part of each preparative SFC platform and must not be neglected. It has been, for a long time, considered as a "drawback" of preparative SFC, necessitating significant investment.

For preparative applications, the amount of needed supercritical CO_2 is obviously much higher than for analytical usage. As a consequence, the bulk CO_2 supply infrastructure need to be adapted. This aspect is often underestimated when setting up preparative SFC platforms and the cost resulting from the installation of large CO_2 delivery unit might rapidly become an important part of the total investment of the SFC laboratory. The most common setups include the utilization of large cryogenic dewar, CO_2 cylinder manifold systems capable of delivering high amount of CO_2 (Figure 2.1A), or the utilization of large tanks (up to several cubic meters) connected to a booster pump (Figure 2.1B). CO_2 delivery might occur directly in the liquid state or as a gas that has been recompressed at the usage location (SFC instrument).

The question of recycling CO_2 is rarely discussed in the literature. Large SFC systems are usually offered or equipped with a CO_2 recycler. It makes certain sense at flow rates above 300 mL/min. At lower flow rates, the recycling option is also available for most instruments but it is rarely applied considering the low cost of CO_2 and the risk of

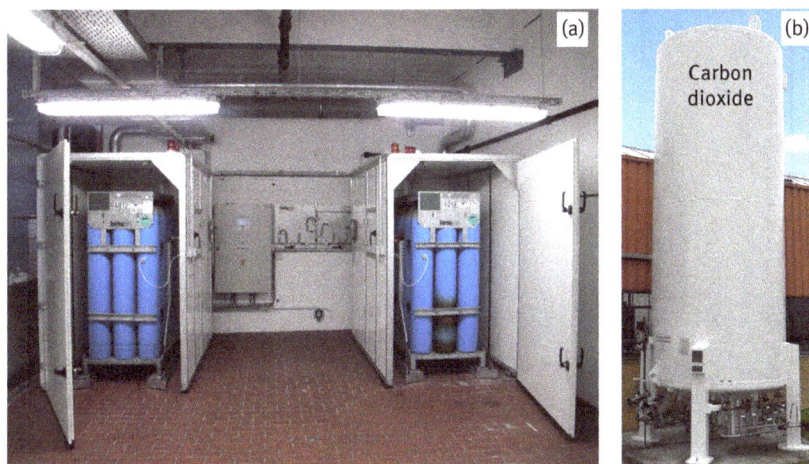

Figure 2.1: Pictures of bulk CO_2 delivery platforms for preparative SFC purifications: (a) Double bundle of CO_2 cylinders in heated cabinets. Each bundle constituted 12 interconnected cylinders. (B) External large single CO_2 tank containing several cubic meters of CO_2.

contamination when dealing with small amount of samples. For large SFC systems (pilot and production), recycling CO_2 is more or less the rule. Majewski et al. from Novasep (France) evaluated the reliability and the impact of CO_2 recycling on solvent consumption.

2.2.2 Available preparative instruments

Basically, the same preparative SFC systems can be used either for achiral puri-fications or chiral separations. While the offer of preparative SFC instruments was very limited 15 years ago, today many manufacturers are offering reliable integrated systems. A comprehensive list of the available preparative instruments and their cha-racteristics is shown in Table 2.3.

The basic design of the preparative instruments is almost identical for all systems. It consists of pumps, injectors, detectors, fraction collection devices, and heaters. The back pressure regulator (BPR) and gas liquid separator (GLS) are further main compo-nents of the system and are specific to SFC instruments.

A schematic diagram of a preparative SFC instrument is shown in Figure 2.2. For preparative applications, particular attention must be paid to injection and fraction collection mode. In addition, the ability to work in stacked injection mode is also an important requirement.

The application types such as sample amount, sample complexity, and through-put requirements are usually determining the instrument choice. The capacity range of the instrument is indicated for each commercially available system and it specifies

Table 2.3: Currently available preparative SFC instruments on the market.

Manufacturer	System	CO$_2$ flow rate range or maximum (mL or g/min).	Modifier flow rate (mL/min)	Optimal column size (I.D. in mm)	Detection (standard or optional)	Fractions	CO$_2$ recycling
Jasco	CSP-4000 Semi-Prep	1–20 mL/min		4–10	UV, PDA, ELSD, MS*, FID, CD	8 or open bed	
	Analytical and semi-prep						
	PR-4088 Prep	150 mL/min	0–50	10–30	UV, PDA, ELSD, MS*, FID, CD	8	
Novasep	Supersep ^{TM} 150	30–150 g/min	0–80	20–30	UV, ELSD	5 + waste	optional
	Supersep ^{TM} 400	100–400 g/min	0–160	30–50	UV, ELSD	5 + waste	standard
	Supersep ^{TM} 1000	250–1,000 g/min	0–400	50–80	UV, ELSD	5 + waste	standard
	Supersep ^{TM} 3000	600–3,000 g/min	0–1000	80–100	UV, ELSD	5 + waste	standard
PicSolution	Hybrid 10–150	150 ml/min	0–50	10–30	UV, MS*	4 + waste	optional
	Analytical and semi-prep						
	PREP 100	100 mL/min	0–50	10–30	UV, MS*	6 or open bed	optional
	PREP 150	150 mL/min	0–50	10–30	UV, MS*	6 or open bed	optional
	PREP 200	200 mL/min	0–100	20–30	UV, MS*	4 + waste	standard
	PREP 400	400 mL/min	0–250	30–76.5	UV	4 + waste	standard
	PREP 600	600 mL/min	0–250	30–76.5	UV	4 + waste	standard
SepiaTec	PREP SFC BASIC	20 mL/min	0–13	4–10	UV, ELSD, MS* (ESI, APCI)	8 or open bed	optional
	Analytical and semi-prep						
	PREP SFC 100	100 mL/min	0–66	10–30	UV, ELSD, MS* (ESI, APCI)	8 or open bed	optional
	PREP SFC 360	360 mL/min	0–240	25–50	UV, ELSD, MS*	8	standard
Waters	Prep 15 SFC	15 mL/min	0–15	4–20	UV, PDA, ELSD, MS*	open bed	No
	Analytical and semi-prep						
	Investigator SFC System	15 mL/min	0–15	4–10	UV, PDA, MS	6	No
	SFC 80q	80 g/min	4–70	19	UV	6 + waste	optional
	SFC 100	100 g/min	5–55	20–30	PDA, MS*	open bed	optional
	SFC 200q	150 g/min	4–70	30	UV	5 + waste	optional
	SFC 350	300 g/min	20–200	50	UV	5 + waste	optional

* optional.

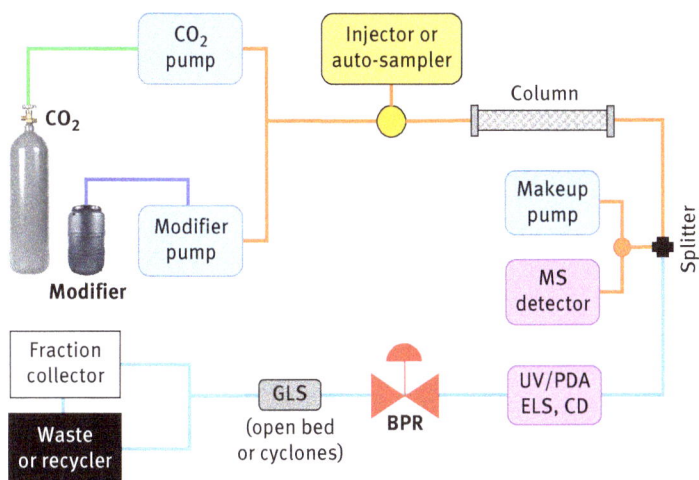

Figure 2.2: Schematic SFC flow path for preparative SFC purification.

the maximum flow rate of supercritical fluid and modifier (Table 2.3). These capacities vary from a few milliliters for systems that offer dual ability (analytical and preparative) and 3 L of CO_2 per minute for large instruments. Small capacity instruments (10–20 mL/min) can be used with usual analytical column to isolate micrograms or a few milligrams of substance and can accommodate columns with inner diameters up to 10 mm. Medium capacity instruments (80–200 mL/min) are usually employed in combination with 15–30 mm i.d. columns. For larger systems (350–3,000 mL/min), columns with inner diameters ranging between 30–50 mm (350–1000 mL/min) and up to 100 mm (3,000 mL/min) have been applied. All systems are equipped with UV/ Photo Diode Array (PDA) detection and optionally Evaporative Light Scattering (ELS) detection. If mass-directed fractionation is required, the choice is more limited but an increasing number of preparative instruments are now offering this option as well.

2.2.3 Injection

Different injection techniques are available and have been applied in preparative SFC. The most common injection modes are mixed stream and modifier stream. A comparison between both approaches has been reported by Miller and Sebastian [28] who found that modifier stream injection generally exhibits a higher performance. A similar comparison was performed by Cox who concluded that there is little difference between the two injection modes [29]. Another new injection method called dual line injection (DLI) has been developed by Jasco [30] that involves incorporating a bypass line in the sample injector of the system. The method is particularly effective for peaks exhibiting shorter retention times. Recently, Shaimi and Cox developed a new injection technique that involves using an extractor to dissolve the dry sample in the supercritical fluid prior to injection. This technique shows good reproducibility and peak shapes [31]. They demonstrated the usefulness of this approach for the preparative isolation of impurities [32].

The influence of injection volume in preparative SFC has also been investigated [33–34]. Rajendran and his group studied the peak distortion phenomenon that is observed when large volumes are injected in SFC. They concluded from their investigation that while modifier-stream injection shows no peak distortion, mixed-stream injection causes severe peak distortion, probably mainly due to the modifier "plug" introduced in the mixed-stream injection [33].

Contrast to HPLC, in preparative SFC it is usual to use rather small columns and inject repetitively small portions of the sample in combination with stacked injections and short cycle times to achieve high productivity. This approach is more or less the rule in SFC, especially for large sample amounts requiring multiple injections. The stacked injection technique consists of injecting the individual small sample solutions before completion of the previous run, to maximize throughput and productivity. The cycle time (time between two injections) is determined on the basis of the retention time of the components and must be optimized for each mixture. Isocratic conditions are applied with stacked injections. This approach is common when operating under isocratic conditions.

For medium- or high-throughput purifications of small samples, single injection is the usual approach and gradient conditions are typically applied. Under these conditions the stacked injection technique cannot be operated.

2.2.4 Fraction collection

While the matter of fraction collection is irrelevant for analytical SFC, it is an important piece of all preparative instruments. The poor sample recovery in preparative SFC has been for a long time a major obstacle for the adoption of the technique. In 1985, Perrut developed the cyclone separator technology for large-scale SFC purification to fix the problem of formation of aerosol during the depressurization of liquid CO_2 [35]. However, the operation of this hardware is relatively heavy and not appropriate for processing many different samples in a short turnaround time. Therefore, for smaller preparative SFC instruments the open-bed format has been preferred. Each manufacturer has developed its own GLS device and fraction collector. The GLS system has an essential function as its design might considerably affect the efficiency of the recovery. Recovery ranging between 80% and 90% are usually achieved with modern preparative SFC–MS instruments [36]. This result is certainly better than what could be achieved with the earlier preparative SFC systems, but still needs some improvements, especially when small amounts of precious material have to be isolated, for example, in medicinal chemistry. In 2012, Bozic presented an in-house designed GLS permitting to achieve > 98% recovery, showing that it is technologically feasible to solve this challenging question [36].

Collected sample solutions in the applied modifier (most usually methanol) are easily evaporated using classical evaporators for large fractions and parallel multi-sample evaporating devices such as Genevac for small sample fractions.

2.2.5 Detection

SFC is compatible with almost all detection methods. Besides PDA (UV) which is the most widespread detection method, mass spectrometry, evaporative light scattering [37–38], polarimetry [39–40], circular dichroism [41], and refractive index [42] have been used in preparative SFC. Coupling MS to preparative SFC instruments has been a great challenge for many instrument manufacturers for more than 10 years. It became commercially available only a few years ago, although earlier works had already demonstrated the feasibility and usefulness of semi-prep SFC–MS 15 years ago [24–26]. They were in-house setup at that time, but MS detector is now available on various preparative SFC instruments and routinely applied in the preparative field, at least at the small and medium scale. As indicated in Table 2.3, an increasing number of preparative SFC instruments are now equipped with MS detectors and MS-triggered fractionation capability.

2.3 Stationary phases for achiral SFC purifications

Earlier applications of achiral purifications by SFC were performed on a very limited number of stationary phases. No SFC-dedicated phases were available at that time and the classical materials such as bare silica gel or C18 silica were employed.

The growing interest for analytical and preparative SFC has considerably changed the landscape and the number of available stationary phases has spectacularly expanded over the last 10 years. This expansion was driven by the need for more efficient and more selective stationary phases permitting to cope with a broad variety of chemical structures, bearing different functional groups (acidic, basic, neutral, small, large, etc.). This feature is important as well for analytical purpose as for preparative purifications.

Figure 2.3 shows the structures of the main achiral stationary phases that are currently available on the market in both analytical and preparative column size. However, this is an instant picture as this market is continuously expanding and is still very dynamic, driven by the growing attraction for SFC analysis and purifications. The picture could be quite different in a few years. The new phases have to be continuously evaluated by the users. However, it is not unlikely that, similarly to what

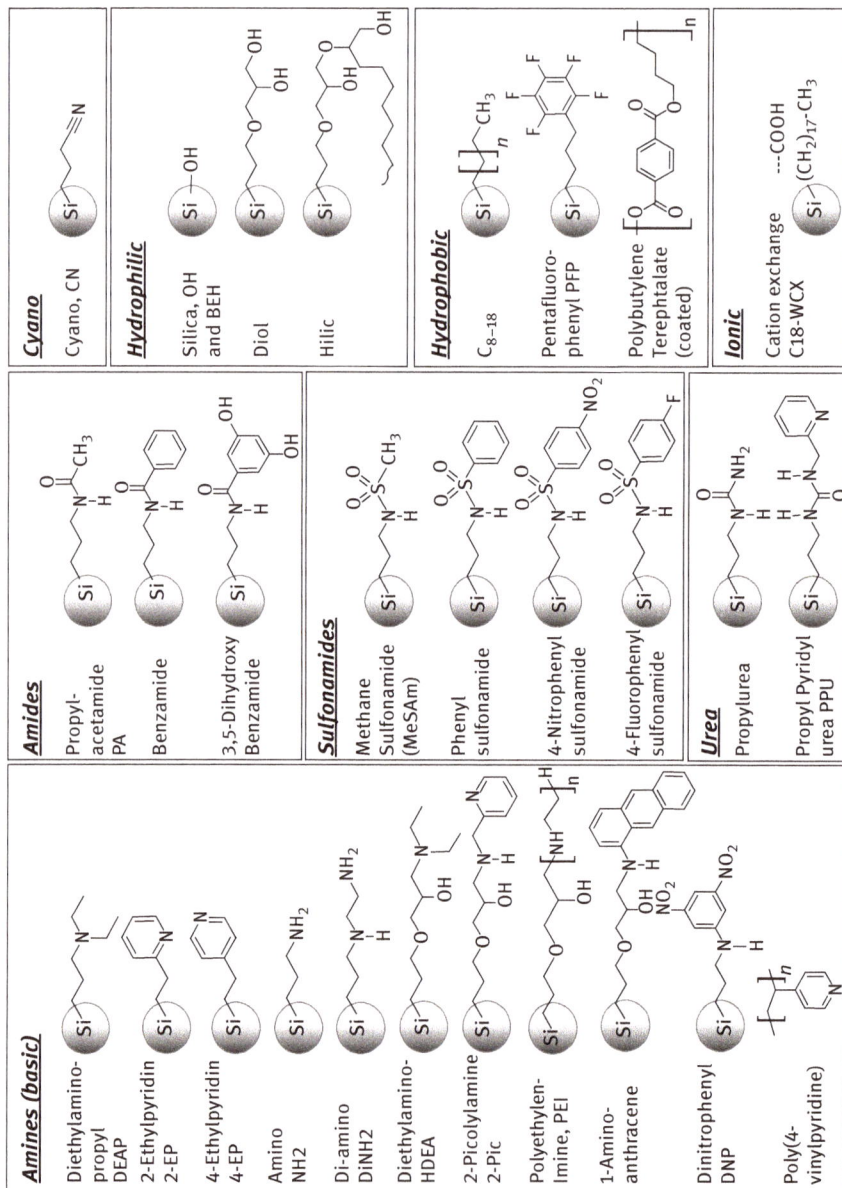

Figure 2.3: Structures of commercially available achiral stationary phases for preparative purification.

occurred in the chiral phase business, few columns chemistries will emerge as the preferred columns in the future.

The phases shown in Figure 2.3 are divided into seven categories according to their functionality for simplification purposes. All achiral stationary phases were obtained by chemical functionalization of silica gel except two polymeric phases that were recently introduced by Daicel Corporation and which were prepared by coating of poly-4(vinylpyridine) and polybutylene terephtalate respectively. Some phases are available from different suppliers and might exhibit different properties, although they have the same basic structure. This is because the fabrication process that varies from supplier to supplier. A number of phases are exclusive and available only from a particular supplier. Among all suppliers, the company Princeton Chromatography offers the widest range of columns.

A much more rigorous classification of the phases has been proposed by Poole [43] and by West and Lesellier [44–48]. West and Lesellier classified the phases according to their polarity, which is a critical parameter for retention, using a quantitative structure–retention relationship (QSRR) and linear solvation energy relationship (LSER) approach. This feature offers a great opportunity to optimize the retention of a particular component of chemical sample mixtures. It is primarily very valuable for analytical method development, requiring robust method for accurate sample analysis and validation. For preparative separations, it might help to identify the most suitable column chemistry and elaborate optimal conditions to purify the target compound to be purified. However, for method development in medium-/high-throughput purifications, it might show some limitations because of the presence of multiple unknown components with various chemical functionalities and basicity/acidity, and showing different polarities. In this instance and considering the high-throughput constraint (short turnaround times), a fast screening of selected phases is probably difficult to avoid.

The chemical properties of the phases shown in Figure 2.3 cover a broad variety of acidity, basicity, and polarities. The importance of the column chemistry on retention time is illustrated in Figure 2.4. Four test compounds bearing different chemical functionalities (basic, neutral, sulfonamide, and acidic) were tested under the same conditions on six column chemistries [23, 49]. This study clearly demonstrates that the affinity of the tested compounds for the different stationary phases considerably varies depending on the column chemistry, leading to changes in the retention time for each particular molecule that is differently affected by column characteristics. The effect of column chemistry is particularly remarkable for the acidic compound Ketoprofen, which is strongly retained on the basic columns "Diamino" and "polyethylene imine (PEI)" (not eluted) and which elutes very early on the "bar silica gel" and Hydrophilic Interaction Liquid Chromatography (HILIC) columns. The opposite effect is observed for the basic compound antipyrine. This mixture of four drugs does not reflect a practical application of a drug mixed with impurities or chemical reactants but nevertheless, it shows the power of the column chemistry as a tool to modulate retention time of the individual components of a complex mixture. This approach implies

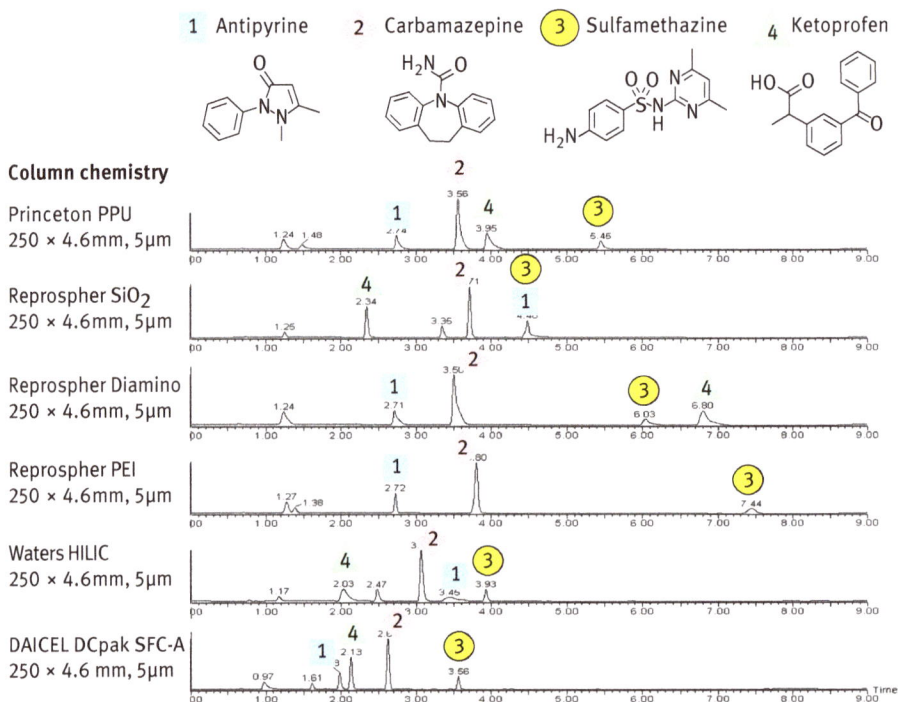

Figure 2.4: Analytical SFC separation of a mixture of four drugs: (1) antipyrine, (2) carbamazepine, (3) sulfamethazine, and (4) ketoprofen on six different column chemistries using a Waters Method Station X5 SFC system. Columns 4.6 mm 250 mm; mobile phase: CO_2/MeOH 90:10; flow-rate, 6 mL/min. Columns (all 5 µm particle size) were made of propylpyridyl urea (PPU) (from Princeton Chromatography), Reprospher SiO_2, diamino, and polyethyleneimine (PEI) (from Dr. Maisch), HILIC (from Waters), and DCpak SFC-A (from Daicel).

the set up of an elaborated method development strategy necessitating an effective screening methodology, similarly to what is commonly applied for chiral separations. The power of applying various column chemistries as a tool to optimize SFC purifications in medicinal chemistry will be discussed in section 2.5 of this chapter.

Anyway, it is interesting to note that, by contrast to reversed-phase HPLC that often utilizes the pH of the mobile phase to modulate retention, in SFC it is the column chemistry (basicity, acidity, lipophilicity, etc.) that is used to modulate retention of the target compound.

It is worthwhile to note that chiral phases, especially those derived from polysaccharide derivatives have also frequently been used for achiral purifications, as they might exhibit particular selectivity. Examples of separations of regioisomers, diastereoisomers, and metabolites have been reported [49–56]. Figure 2.5 shows the preparative SFC separation of two hydroxylated metabolites produced by microbial biotransformation [23, 49].

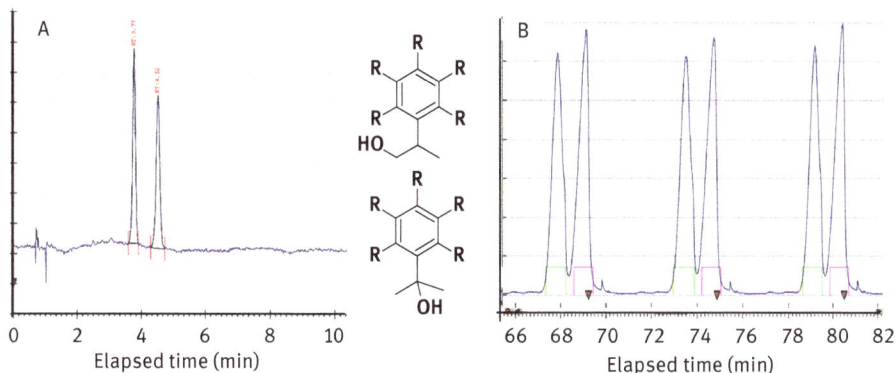

Figure 2.5: Semi-preparative purification of hydroxylated metabolites. Column: Princeton DEAP, 4.6 mm × 250 mm (5 μm); Mobile phase: CO_2/MeOH 90:10, isocratic; Flow rate: 6 mL/min. (a) Analytical separation. (b) Semi-preparative separation. Stacked injections (3 cycles shown); Cycle time, 5.7 min (18 injections).

2.4 Applications of preparative achiral SFC purifications on packed columns

2.4.1 Small- and medium-scale applications

The potential of SFC for preparative purification of chemical substances and materials had been recognized since a long time before it was applied to chiral separations. Preparative achiral SFC purifications have been around for more than 30 years, but were mostly applied to lipophilic compounds. For example, Hartmann and Klesper already reported the preparative separation of styrene oligomers by SFC in 1977 [57].

Table 2.4 lists most achiral preparative SFC applications that were published from the early beginning of SFC until about 2015. This does not mean that the list is exhaustive, as further applications have very likely been performed in the industry but were not disclosed for intellectual property reasons. This assumption is corroborated by the fact that an appreciable number of presentations given in the framework of the annual Conference on Packed Column SFC (since 2007) reported achiral purifications but without disclosing the sample structures.

The applications in Table 2.4 are sorted according to the investigation fields and should give a flavor of what kind of molecule could be purified by SFC.

There were ongoing activities in the field of preparative SFC purifications, particularly in France at the beginning of the 1980s [15, 58–59]. The major focus at that time was the development of more efficient and competitive processes for large-scale technical applications. Isolation of tocopherol from wheat germ oil [60] and

Table 2.4: Preparative achiral SFC applications.

Substances/Purpose	Structure	Stationary phase	Purified amount	Reference
Polymers, olygomers, and hydrocarbons				
Heavy hydrocarbons mixtures, naphthalenes *Separation*	(naphthalene with R^1, R^2 substituents)	Silica gel C_{18} silica	n.d.a.	[35]
Polystyrene oligomers *Characterization*	$-[CH-CH_2]_n-$ (phenyl)	n.d.a.	6–20 mg	[57, 83]
Paraffin *Characterization*	$H_3C-[CH_2]_n-CH_3$	n.d.a.	n.d.a.	[84]
Carbosilanes and phosphines *Purification*	(silane structure) $R = (CH_2)_n CH_3$	C_{18} silica	n.d.a.	[63]
Fullerene *Characterization*	(fullerene structure)	Silica gel	2.4 mg	[85]
Poly(methyl methacrylate) Poly(chloral), Oligo (oxymethylene) diacetate *Characterization*	(polymer structure)	n.d.a.	n.d.a.	[86]

Compound	Structure	Stationary phase	Amount	Ref.
Poly(methyl methacrylate) oligomers *Characterization*		Silica gel	50 mg	[62, 87–89]
δ-valerolactone and ε-caprolactone oligomers *Characterization*		Silica gel	12 mg/inj.	[66]
Propylene oxide oligomers *Characterization*	$CH_3OCH_2CH(CH_3)OCH_2CH(CH_3)OCH_3$ $CH_3OCH_2CH(CH_3)OCH(CH_3)CH_2OCH_3$ $CH_3OCH(CH_3)CH_2OCH_2CH(CH_3)OCH_3$	Silica gel	n.d.a.	[90]
Poly(ethylene glycol) oligomers*Characterization*	$H_3C-[CH_2CH_2-O-]_n CH_3$	Silica gel	n.d.a.	[91]

Drugs and drug Intermediates

Compound	Structure	Stationary phase	Amount	Ref.
Steroid hormones: Cyproterone acetate (antiandrogen)		Silica gel	8 mg	[64]
17α-estradiol 17β-estradiol		Nitrophenyl	2.5 mg	[65]
Drug intermediate		2–Ethylpyridine	43 g	[92]
Drug intermediate		Chiralpak IC	10 g	[49, 50, 92, 93]

(continued)

Table 2.4 (Continued)

Substances/*Purpose*	Structure	Stationary phase	Purified amount	Reference
Drug intermediates		Chiralcel OD-H 2–Ethylpyridine Diol	3–5 g	[56]
Sulfamethazine Sulfaquinoxaline Sulfadimethoxine (veterinary drugs)		Amino	mg amounts	[94]
Drug intermediate	Structure not disclosed	Cyano	425 g	[95]

Drug impurities and metabolites

| Sinalbin degradation products *Characterization* | | Silica gel | 2 mg/inj. | [81] |
| Atorvastatin impurities *Characterization* | | Pyridine amide (PPA) | mg amounts | [82] |

Metabolites *Characterization*	Diethylamino pyridine (DEAP)	15 mg	[51]
Metabolites *Toxicity study*	Ethypyridine	n.d.a.	[139]

Natural products

Fish oil: Docosahexaenoic acid (DHA), Ecosapentaenoic acid (EPA). *Food additive*	Silica gel C$_{18}$ silica Perfluorophenyl Nitro Aminophenyl	350 mg to tons	[96–102]
Tocopherolalpha, beta *Food additive*	Silica gel	100 mg	[60, 103]
Phytol isomers *Food additive*	Silica gel	Large amounts of SMB–SFC	[104, 105]
Phytol isomers α-, γ-, δ-Tocopherol *Food additive*	Silica C$_{18}$	10–65 mg	[115]

(continued)

Table 2.4 (Continued)

Substances/*Purpose*	Structure	Stationary phase	Purified amount	Reference
Lemon-peel oil: α-Pinene, β-Pinene Limonene, γ-Terpinene Nonanal, Neral, Geranial *Flavors*		Silica gel	4 g	[61]
Kava Lactones		Protein C4	n.d.a.	[106]
Vitamin D3Previtamin D3 *Food additive*		n.d.a.	n.d.a.	[107]
Ferulate-phytosterol esters (cholesterol lowering agent) *Food additive*		Amino	6 g	[108]

Rosemary constituents *Antioxidant*	Caffeic acid; Vanillic acid; Carnosic acid; Carnosol; Rosmarinic acid	Diol Phenylsilicone 2–Ethylpyridin	n.d.a.	[109–111]
Natural products: Methyl asterrate Asterric acid		Diol	72 mg	[71]
Natural products: Trichurusin J Trichurusin K		Diol	16 mg	[71]
Natural products: Ascosalitoxin		Pyridine-amid	n.d.a.	[71]

(continued)

Table 2.4 (Continued)

Substances/*Purpose*	Structure	Stationary phase	Purified amount	Reference
Natural products: Memnobotrin isomers		Ethypyridine	9 mg/inj.	[71]
Natural products: Aurovertin mixture		Amino (NH₂)	153 mg	[71]
Natural products: Staplabin		Pyridine	192 mg	[71]
Natural products: Antibiotic X 14881A		Amino (NH₂)	449 mg	[71]
Curcumin DiMeO curcumin Bis dimethoxy curcumin		Viridis BEH OBD	45 mg	[72]

Natural Products : Psoralen Bergapten Xanthotoxin Isopimpinellin		Silica gelC₁₈ silica Perfluorophenyl Nitro Aminophenyl	n.d.a.	[74]
Natural products: Polymethoxyflavone Tangeretin Nobiletin		Chiralpak AD	5 g	[75–76]
Palm oils constituents: Carotenes Vitamin E (Tocopherol) Sterols		Silica gel and Silica C₁₈	n.d.a.	[112]
Squalene *Food additives*				
Sucrose ester from Tobaccos *Characterization*		n.d.a.	n.d.a.	[113]

(continued)

Table 2.4 (Continued)

Substances/Purpose	Structure	Stationary phase	Purified amount	Reference
Citric acid *Food additive*		Cyano	n.d.a.	[114]
Artemisinin Anti-malarial agent *Drug*		Silica C$_{18}$	n.d.a.	[116]
Resveratrol Emodin		Silica C$_{18}$	Large amounts of SMB–SFC	[117]
Peptides, Cyclopeptides				
Cyclopeptide epimers		4–Ethylpyridine	50 mg	[49]
Hexapeptide epimers		Chiralpak IA	20 mg	[92]

Compound	Structure	Stationary phase	Amount	Ref.
Cyclosporin A *Drug*		Silica gel	20 g/kg silica	[118, 119]
Pesticides				
Piperonyl butoxide *Insecticide*	$CH_2O(CH_2)_2O(CH_2)_2OC_4H_9$	Silica C_{18}	53 mg/ inj.	[120]
Miscellaneous				
Glucoide anomers		Silica gel	10 mg	[78]
[11]C-Labeled compounds: [methyl-^{11}C]anisole, L-[methyl-^{11}C]methionine 4–methoxyphenyl[^{11}C] guanidine		SFC in ammonia	n.d.a	[70]
Additives in PVC *Characterization*	Not available	Silica gel	n.d.a	[121]

Note: n.d.a: no data available.

fractionation of lemon-peel oil [61] by SFC were described in 1990. Another favored application field during the same period of time has been around polymers and oligomers such as methyl methacrylate oligomers [62] or functional silanes and phosphines [63]. Most applications were clearly directed to lipophilic substances as it was considered that the very weak polarity of CO_2 would be appropriate for the purification of apolar compounds. Almost all earlier applications were performed using pure carbon dioxide.

Between 1990 and 1999, the interest in using preparative SFC for small-scale applications was relatively limited, very likely because of the technical shortages (pressure regulation, injection, aerosol formation, etc.). Nevertheless, a number of applications for purification of achiral mixtures have been performed at various scales, ranging from milligrams to hundreds of grams. Different types of molecules were purified, comprising the isolation of steroid hormones [64–65], synthetic lactone oligomers [66], and isolation of flavor and food constituents [67]. More than 20 years ago, Rocca et al. already pointed out the potential of preparative SFC to isolate and characterize small impurities in chemical mixtures [68], comparing the efficiency of both preparative LC and SFC. There was also a report on the preparative SFC of an ^{11}C-labeled compound using ammonia as the supercritical fluid [69].

Since 1997, there was a growing interest for the SFC purification of docosahexaenoic acid (DHA) from fish oil [70]. DHA is an omega-3 fatty acid, which has an essential biological function in human brain and other organs.

From the late 1990s, more robust instruments and new stationary phases became available. Although the interest of the "new" users has primarily focused on the potential of preparative SFC for chiral separations, the number of achiral preparative application has continuously grown. Most recent applications include the purification of synthetic intermediates, the isolation of natural products [71–76], the purification of peptides [77] and cyclopeptides, the separation of anomeric derivatized glucosides [78], and the isolation of metabolites [49]. The practical interest of the utilization of preparative SFC for the isolation of drug-related impurities for identification has been reviewed in two papers dating from the end of the last decade [79–80]. Preparative SFC was also found to be useful for the isolation and identification of drug degradation products [81–82].

Based on the published works of the last 20 years, it seems that natural products dominate in terms of number of preparative SFC (Table 2.4).

Most of them have been used as drug or food additives. However, as mentioned earlier in introduction of section 2.4.1, this observation must be taken with caution as the published material does not necessarily reflect the actual picture of the utilization of preparative SFC in the industry that has been the major user. Figure 2.6 shows the preparative SFC separation of two isomers of an antibiotic X 14881A sample produced from an actinomycete strain. The purification was performed isocratically on an amino column using the stacked injection method. Both isomers were isolated in good yield and high purity.

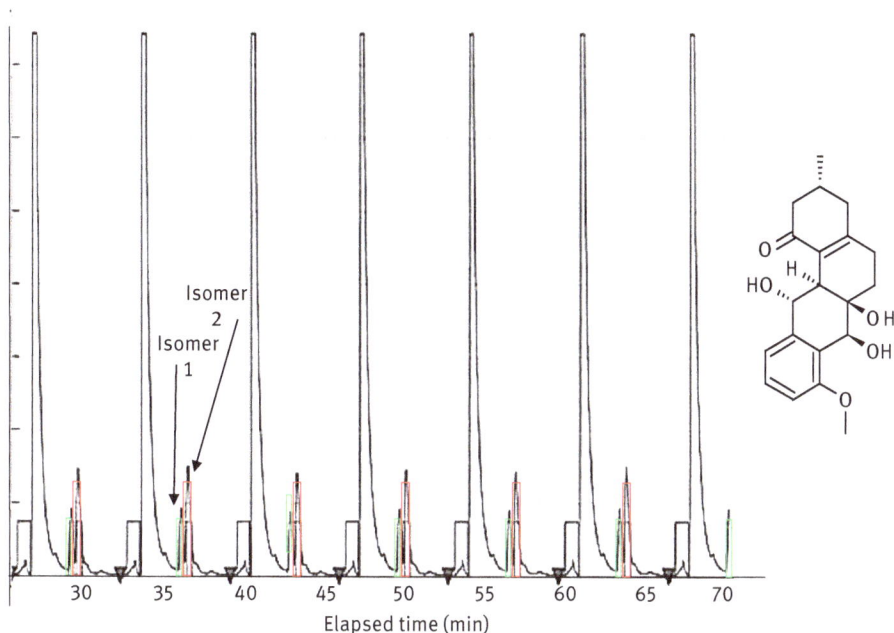

Figure 2.6: Preparative SFC separation of 449 mg of an isomer mixture of an antibiotic (X 14881A) sample from an actinomycete strain. Column: Reprospher Amino, 30 mm × 250 mm (5 μm); mobile phase: 10% EtOH in CO_2, isocratic; stacked injections, 7 cycles.

Most of the published applications are small or modest in terms of amounts (Table 2.4). However, there are probably more applications in the range of 100 g to kilograms but they have not been published because of confidentiality reasons. Figure 2.7 shows one of such a preparative example of substance (structure not disclosed) that has been purified at the kilogram scale. Forty stacked injections of 2.1 g with a cycle time of 1.35 min. on a cyano column are displayed on the chromatogram [95].

2.4.2 Flash supercritical fluid chromatography

Flash chromatography is a well-established technique for purification of chemical substances, intermediates, and final products in synthetic chemistry laboratories. Almost all laboratories in academic and industrial research organizations are applying this technique all over the world on a daily basis. Flash chromatography is a normal phase chromatography mode, employing organic solvents as the mobile phase. As such and considering the popularity of this technology, it consumes (cumulated) huge amounts of organic solvents, which are usually eliminated by

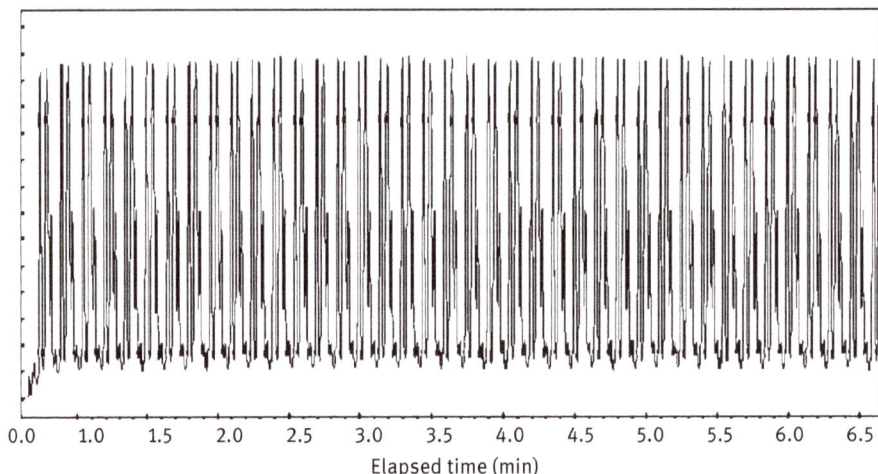

Figure 2.7: Preparative SFC purification of 425 g of a drug intermediate. Column: Princeton Chromatography Cyano, 50 mm × 250 mm (5 µm); mobile phase: 25% MeOH in CO_2, isocratic; flow rate, 250 mL/min; temperature 35 °C; stacked injections (cycle time 1.35 min); injection, 2.15 g/inj. (Adapted from [95]).

burning. Therefore, there is a high solvent and cost-saving potential in developing an SFC version of liquid flash chromatography. There are many technical challenges associated with this development. First of all, the equipment needs to be stable, robust, safe, and simple to use (user-friendly). The challenges also include the question of method development (replacement for thin layer chromatography [TLC]), and the substantial investments for large-scale CO_2 delivery platforms. Miller investigated the possibility to use TLC data as a basis for SFC separations but he could not find direct and systematic correlation between retention in LC and SFC for a series of tested compounds [122]. Independent of the SFC method development challenge, Miller and Mahoney showed that flash SFC could achieve similar loadings as flash LC [123]. More recently, Ashraf-Khorassani et al. published a paper demonstrating the feasibility of correlating TLC separation of ternary mixtures of neutral analytes with SFC [124].

F. Riley at Pfizer [125] reported preliminary successful results in developing a flash SFC device in 2008. In 2014, a flash SFC prototype was presented at the SFC conference in Basel [126]. This prototype has been tested more deeply by McClain et al. [127] but this promising apparatus has not yet concretized in a commercial instrument. Another flash SFC instrumentation has been developed by "Modular SFC" but no performance details are available [128].

If SFC continues to attract more attention, there is probably a future for this kind of technique, but it will still need some efforts to convince chemists to switch from flash HPLC to flash SFC.

2.4.3 Large-scale and industrial applications

Even though SFC has been successfully implemented in number of companies up to pilot scale to separate enantiomers at the kilogram scale, it has not yet been really embraced as a purification technique in production. It might be that some larger preparative SFC units are implemented in the industry but no information is available.

Only a few semi-industrial applications have been reported. Among them, purification of high value compounds DHA and eicosapentaenoic acid (EPA) is probably the most relevant application in the literature (Table 2.4).

The interest in the separation of these valuable components started in the mid-1990s and a pilot unit was built in Japan in 1997 [97]. DHA is an omega-3 fatty acid that has an essential biological function in human brain and other organs and has a high commercial value. The technical and economic feasibility of separating DHA-ethyl ester and EPA-ethyl ester from tuna oil using SFC was studied in details by Akio et al. [98]. Lembke developed a production unit using a column of 1.15 m by 10 cm i.d. and running at a flow rate of 200 kg/h [101]. This process permitted to reduce the production costs by 75% compared to the HPLC method. Further examples of large-scale SFC purifications include purification of the insecticide piperonyl butoxide [120] and the separation of cis and trans phytol [104]. Productivity of 1.42 kg/day of purified piperonyl butoxide could be reached using a 25 cm x 10 cm i.d. column (C18) at a flow rate of 90 kg CO_2/h, cycle time 2 min.

Separation of phytonutrients (carotenes, vitamin E, sterols, squalene, α-tocopherol, α-tocotrienol, γ-tocopherol, and γ-tocotrienol) from palm oil at a pilot scale has been reported at the annual Conference on Packed Column SFC 2008 in Zurich (Table 2.4) [112]. A SFC 600 unit was used with a 35 cm I.D. column at a flow rate of 600 kg CO_2/h). The details of the process have recently been reviewed [129].

2.4.4 Preparative SFC by simulated moving chromatography (SMB)

As a logical consequence of the successful development of SMB in liquid chromatography, in particular for chiral separations, an SFC version of this continuous chromatographic technique has been developed by two groups: (1) Novasep in France, which is a leading company in the field of large chromatographic equipment [130] and a few years later, in (2) Germany at the Technical University of Hamburg [105]. The feasibility of performing supercritical fluid SMB (SF-SMB) was demonstrated by both groups but the interest remained very low, considering the technical challenge and the fact that the benefit was insignificant compared to LC-SMB in terms of greenness, as the latter technique is highly efficient regarding solvent consumption and solvent recycling (up to 99.5%). In 2008, Novasep presented an improved version of the previously developed SF-SMB system, stating that the new design permits to considerably improve the productivity of the process [131].

Comparison between batch SFC and SF-SMB has been performed and the authors concluded that batch SFC chromatography affords a higher productivity [132–133]. Nevertheless, recently there was a regain of interest for SF-SMB at the Shou University in Taiwan in the group of Liang. He developed SF-SMB processes for the separation of various substances of pharmaceutical interest such as sesamin and sesamolin [134] or resveratrol and emodin [117].

2.5 Achiral preparative SFC purifications in medicinal chemistry

The interest of research scientists in utilizing SFC as a tool to purify more polar molecules, in particular drug and drug intermediates, on a broader basis and as a complementary technique to the popular reversed-phase HPLC is relatively recent. In 2004, Searle and coworkers compared the performance of preparative HPLC/MS and preparative SFC for the high-throughput purification of synthetic libraries [135]. At that time, MS was not available on commercial preparative SFC systems. They concluded that both techniques HPLC and SFC gave similar results in terms of purity and recovery with few exceptions. Although the preparative SFC system proved to be less robust for high-throughput operations and the lack of MS-triggered fractionation capability was a clear drawback compared to HPLC, it was already emphasized that SFC had a great potential. In 2005, White and Burnett compared reversed HPLC and SFC to separate a set of drugs. Based on these successful results, they proposed a strategy for achiral batch purification of drugs by SFC. However, they tested only one column chemistry (2–ethylpyridine) [136]. In the meantime, a broad variety of new column chemistries have appeared on the market for achiral analysis. The new phases have permitted to considerably improve the separation performance and extend the applicability of SFC changing the strategies for method development.

The breakthrough came in 2008 with the introduction on the market of the first preparative SFC instrument equipped with mass-directed fractionation [27]. Coincidentally, there was a severe issue because of a worldwide acetonitrile shortage during the same period of time [13–14], motivating the chromatographers to explore or rediscover alternative purification techniques to replace RP-HPLC. Moreover, the awareness for environmental considerations had considerably increased since the beginning of the millennium. For years, it has been a real challenge to develop a robust and reliable preparative SFC-MS instrument, offering mass-directed fractionation. This has been a considerable obstacle to the development of preparative SFC for achiral purification of complex mixtures for which mass-directed fractionation is a must. A few research groups had successfully developed their own SFC-MS purification system, in particular in the context of combinatorial chemistry projects [24–26], but these instruments were not accessible to the global community of separation scientists and limited in their applicability with respect to sample size.

With the introduction of a dedicated preparative SFC-MS instrument, things have drastically changed. Several groups evaluated more or less simultaneously the suitability of the new instrument to replace, at least partially, the "unmatched" reversed-phase purification technique in medicinal chemistry by the SFC technique, although these two chromatographic modes are totally different [51, 137–146]. All these groups essentially came to the same conclusion and demonstrated that the purification of small synthetic libraries (sample mixtures) by preparative reversed-phase HPLC-MS and a SFC-MS globally show equal performance with respect to purity and recovery, but, that on average SFC was three times faster than RP-HPLC. In 2012, Francotte colleagues reported that thousands of samples were purified annually using this approach in the drug discovery department of Novartis and that about 80% of the submitted samples could be purified by preparative SFC-MS [147]. Recently, Rosse at Dart Neurosciences developed a platform to purify annually more than 90,000 small size samples using the same MS-triggered preparative SFC instrument and applying a generic method [148]. This successful and rapid expansion of achiral preparative SFC was because of the concomitant development of appropriate SFC instruments and of new stationary phases.

Today, most large pharmaceutical companies have their SFC platform for purification on medicinal chemistry samples and it can be estimated that, cumulated for all companies, *hundreds of thousands* of medicinal chemistry samples are purified by SFC.

2.5.1 Method development for medium throughput preparative achiral SFC

Originally, high-throughput purification was defining a process aiming to purify a high number of chemical samples within a very short time. In order to meet these requirements, a generic purification method is generally applied. The method implies the selection of defined chromatographic conditions (column, gradient, temperature, and run time), which are supposed to be suitable for a majority of samples. Under these conditions, it is generally accepted to "lose" some samples and to compromise on quality and yield of the purified target compound. This was the case in the field of purification of large synthetic libraries (combinatorial chemistry). This compromise is not acceptable for final compounds to be submitted to biological testing in drug discovery. These samples require full attention as they are generally obtained from multiple and complex synthetic steps. Recovery and purity are of most important in this case and these constraints necessitate the development of a "tailor-made" purification approach for each individual sample. Since many years, most of these purifications were performed by MS-triggered reversed-phase HPLC, but following the introduction of MS-triggered SFC, a rapidly growing number of users moved to SFC, motivated by the numerous advantages of the technique (Table 2.2).

As discussed earlier in this chapter (Section 2.3), a broad range of column chemistries are now available for preparative SFC on the market and this variety permits to modulate and optimize the chromatographic conditions for each particular sample, regarding retention, selectivity, and peak shape. This situation has led the set up of strategic approaches for method development in achiral preparative SFC, similarly to the approaches that have been elaborated for chiral separations.

A systematic evaluation of achiral stationary phases has been performed by a few groups [149–152], but the column chemistry field is extremely dynamic and rapidly changing with new stationary phases being regularly launched. Moreover, these earlier evaluations were generally performed with a mixture of selected "model compounds" that are supposed to reflect the variety of molecular structures. In reality, mixtures to be purified generally consist of structurally related substances generated during the nonselective chemical reactions.

Based on the purification of tens of thousands of chemical reaction mixtures, Francotte and his group reported that practical experience indicates that some columns are more suitable for basic and others for acidic compounds, and that some accommodate both types of molecules [23, 49]. They also reported that a combination of 3–4 columns allows the purification of more than 90% of the mixtures submitted by medicinal chemists [92]. The four phases are the basic phase PEI, the ionic phase C18-WCX, the HILIC silica phase, and the propylpyridyl urea (PPU) phase (Figure 2.3). The 2-ethylpyridine, 4-ethylpyridine, and diamino phases might also be very useful for some classes of substances. In conclusion, experience showed that the column chemistry (basicity, acidity, and lipophilicity) can be exploited to modulate retention of the target compound in SFC.

Each pharmaceutical company has developed its own general workflow for preparative SFC-MS purification, but they all basically follow the same scheme. The workflow that we apply for small size samples in drug discovery is probably representative [23, 49]. This workflow is outlined in Figure 2.8.

In the presented approach, the submitted sample, typically ranging between 50 and 200 mg, is subjected to an analytical screening on various column chemistries. As the retention of the single components of a mixture on a defined stationary phase cannot be predicted, it is important to rely on an efficient screening platform to maximize the quality of the separation with respect to the target substance. The molecular mass of the molecule being critical information for selecting the optimal column to purify the target compound, the analytical screening must also be performed by SFC-MS. This can be easily achieved, using a commercially available SFC-MS system equipped with 5 columns in parallel. When the bank of preferred 5 columns is not successful, there is the option to switch to a second or third bank of 5 columns. After identification of the best column chemistry, a validation run is performed to collect the data needed to determine

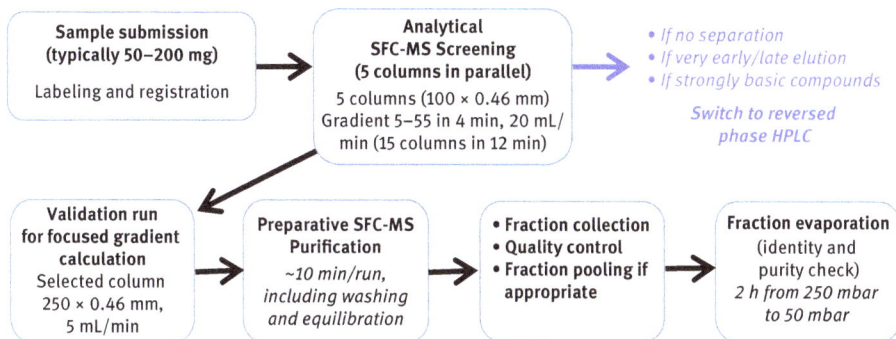

Figure 2.8: General workflow for achiral MS-triggered SFC purifications.

the optimal focused gradient for the preparative purification. This gradient calculation is necessary because the flow in each individual column of the parallel screening unit is not necessarily equal, as there is no flow control on the currently available SFC-MS parallel screening unit. The focused gradient calculation procedure has been empirically determined and is specific for each column type and size and for a specific instrument [49]. After identification of optimal chromatographic conditions, the preparative purification is executed and fractions containing the target compound(s) are collected. Purity control of the fraction can be performed by SFC or RP-HPLC (orthogonal method) or both. The purified substances can be delivered as concentrated solution in the used modifier or as dried material after evaporation of the residual organic solvent. This present setup permits to reach short turnaround time (maximum 24 h after submission) and high purity samples that are critical in drug discovery.

In most cases, one single injection is sufficient for processing the submitted sample. If more substance is submitted (> 100 mg), multiple injections are usually performed. Stacked injections are not feasible due to gradient conditions.

In the presentation [49], it was also emphasized that, especially for preparative applications, the use of additive should be avoided as far as possible. Indeed, additives can be the source of multiple issues such as difficulty in removal, chemical reactions (decomposition) with acidic or basic sensitive molecules, and salt formation with basic substrates. Particularly, TFA and TFA salts that are commonly present in samples purified by RP-HPLC are undesirable in biological assays because of cell toxicity.

In terms of preparative SFC applicability, it has been shown that while SFC is suitable for the purification of most substances having a pKa below about 8.2, it is less appropriate for those substances exhibiting a pKa higher than 8.2 (Figure 2.9) [49]. For these substances, long retention times or even no elution, poor peak shape, and broad tailing are generally observed independently of the used stationary phase. In that instance, reversed-phase is probably a better option. The ClogP value of the substances does not seem to significantly affect the SFC applicability. The study was achieved with a set of 120 "real" samples that were tested on 5 to 10

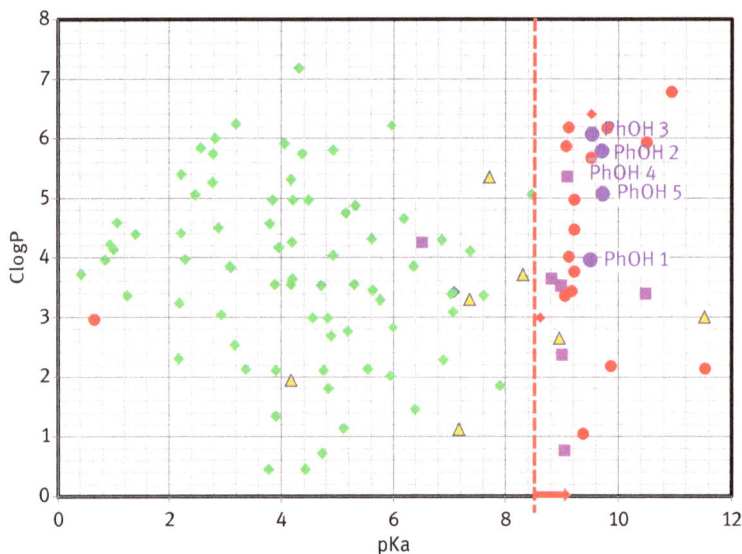

Figure 2.9: SFC applicability vs. ClogP and pKa of target compound.

different column chemistries (2–ethylpyridine, 4–ethylpyridine, amino, di-amino, diol, silica, propylpyridyl urea, diethylaminopropyl, cyano, and perfluorophenyl). Green spots in Figure 2.9 correspond to substances exhibiting reasonable retention times and good peak shape (SFC successfully applicable). Yellow spots correspond to substances exhibiting a relatively long retention time and/or an irregular peak shape on most of the tested phases but SFC is successfully applicable at least on one column. Violet spots correspond to substances exhibiting a long retention time and/or a poor peak shape on most of the tested phases but SFC is successfully applicable at least on one column. Red spots correspond to substances exhibiting a long retention time (or not eluted) and a poor peak shape on all the tested phases (SFC not recommended).

2.5.2 Applications of preparative achiral SFC in medicinal chemistry

It can be estimated that several hundreds of thousands samples are yearly purified by SFC worldwide in the pharmaceutical industry today. The technique can be applied to almost all kind of small molecule drugs including small peptides and cyclopeptides. There are only a few exceptions comprising strongly basic molecules and compounds that are insoluble in common organic solvents. In spite of the vast number of applications, only very few have been published because of intellectual property reasons.

Figures 2.10–2.12 illustrate the discussed workflow (Figure 2.8) for sample mixtures containing acidic and basic target substances respectively. They are typical applications of achiral SFC purifications of medicinal chemistry sample mixtures [49]. Figure 2.10 shows the results of screening of a mixture containing the target substance (carboxylic acid molecule) on five different column chemistries. The picture distinctly shows the influence of the column chemistry on the elution pattern of the various components of the mixture and in particular on the peak of interest. A late elution is observed on the diethylaminopropyl and the 4-ethylpyridine phases while the peak of interest partially co-elutes with a minor impurity on the silica gel and the diol columns. In this example, the 2-ethylpyridine column provides the best chromatographic elution pattern.

Figure 2.11 shows the results of the screening of a crude synthetic mixture containing three target substances (basic compounds) on five different column chemistries. The chromatogram traces exhibit a poor separation of the individual target

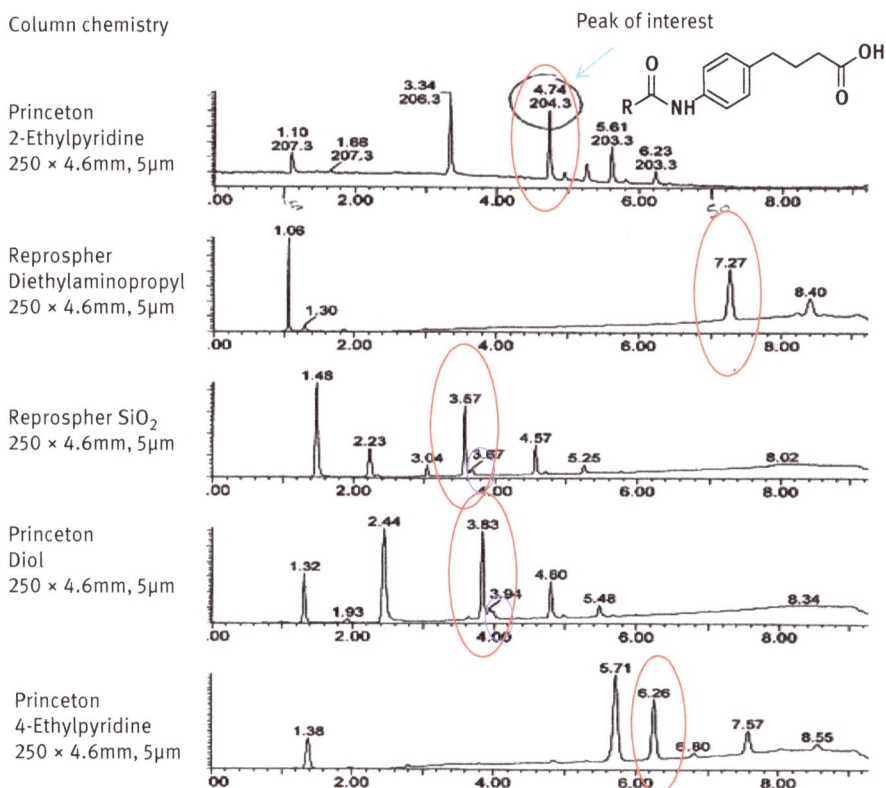

Figure 2.10: Analytical SFC screening of a typical medicinal chemistry sample (crude synthetic mixture) containing a carboxylic acid molecule as the target substance on five different column chemistries. Target compound is circled in red; columns 250 mm × 4.6 mm; flow rate 20 mL/min; gradient mode 5–50% MeOH in CO_2 within 6 min.

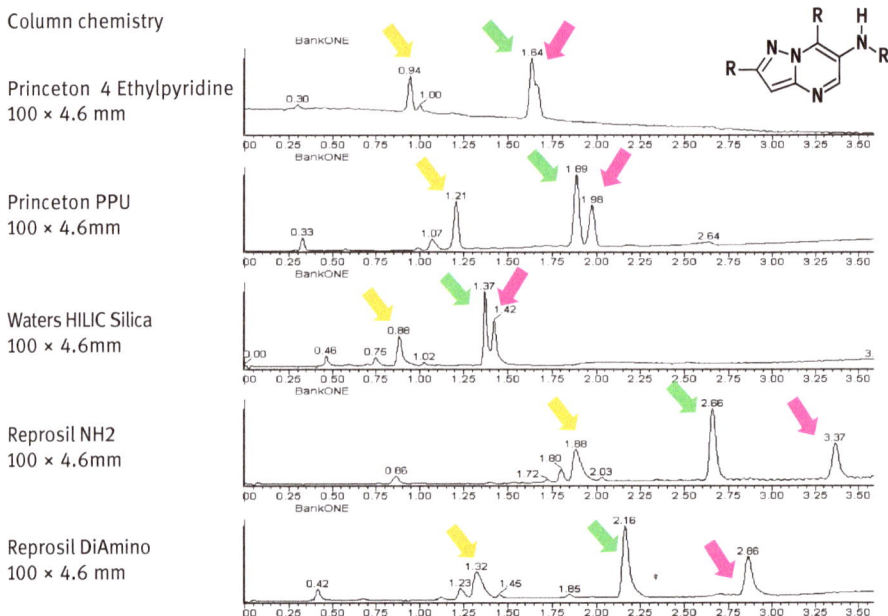

Figure 2.11: Analytical SFC screening of a medicinal chemistry sample (crude synthetic mixture) containing three compounds of interest (basic substances). Method development using SFC-MS Waters X5 station (5 columns screened in parallel). Columns 250 mm 4.6 mm; flow rate 20 mL/min; gradient mode 5–50% MeOH in CO_2 within 6 min.

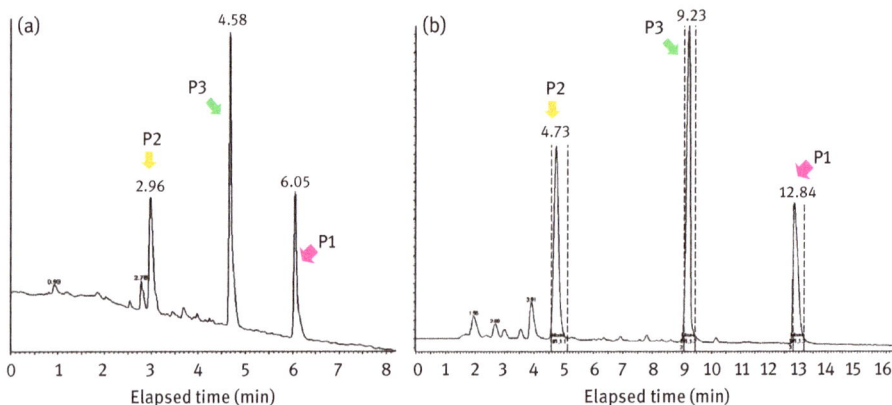

Figure 2.12: (a) Validation of analytic run on selected column (Reprosil DiAmino, 250 mm × 4.6 mm, 5 μm); (b) SFC-MS preparative purification using Thar-Waters Prep 100 SFC-MS; column, Reprosil Diamino, 250 mm × 30 mm, 5 um; 70 mg dissolved in 2mL MeOH/dichloromethane 1/1; Gradient 10 to 48% MeOH within 17 min; flow-rate 100 mL/min.

compounds on the 4-ethylpyridine, PPU, and HILIC columns (peak overlapping). The retention time of the last eluted peak on the amino column is relatively long. For this mixture, the diamino column provides the best chromatographic conditions in terms of selectivity and retention time for the three peaks of interest.

Figure 2.12 shows the chromatogram of the analytical validation run obtained on the analytical diamino column (Figure 2.12A) and the chromatogram of the correspon-ding preparative run on a larger column (Figure 2.12B). The correspondence between the analytical and preparative runs is clearly visible.

In many pharmaceutical companies, achiral SFC has become the first technology choice for the purification of small molecules (final products), replacing reversed HPLC in about 80% of the samples. The application range is continuously expan-ding and covers drug molecules, natural products, metabolites, and small linear or cyclopeptides. There is also a growing demand for the purification of synthetic intermediates.

This shift from RP-HPLC to SFC can be considered as a "green" switch as SFC consumes on average about 20% less organic solvents. Moreover, in RP-HPLC prepara-tive applications, the removal of water is an energy-consuming process necessitating about 7 times more energy compared to the SFC sample evaporation. Additionally, it permits to considerably reduce the use of acetonitrile, which is highly toxic by inhala-tion and penetration in the body through the skin.

It might be anticipated that the number of application of preparative SFC will continue to grow in volume and in variety. RP HPLC will still be needed for very basic and polar compounds but the impact of this chromatographic mode for preparative purifications of small molecules (final products and intermediates) will very likely considerably diminish.

2.5.3 High-throughput achiral SFC purifications

In high-throughput synthesis, a high number of samples are generally generated, usually in small amounts, and these numerous sample mixtures have to be puri-fied. Given these requirements, it was consistent that some scientists were looking for a fast purification technique. In this context, utilization of preparative SFC for high-throughput purification started to attract a number of groups involved in com-binatorial chemistry. Already 18 years ago, Coleman proposed a protocol for high-throughput purifications of samples from combinatorial chemistry libraries using SFC and applying generic conditions [153]. High-purity substances could be rapidly isolated up to the tens of milligram levels. One year later, Berger et al. published the development of a new separator for automated fraction collection in semi-prepara-tive SFC (up to 50 mg of solute) permitting to avoid the problem of aerosol formation [154]. Kassel and his team improved the system by incorporating mass-directed frac-tionation [20]. Further groups developed high-throughput SFC purification platforms with UV-triggered fraction collection [155–157]. In almost all cases, purifications were performed in the context of hit finding activities where large libraries of potentially new drugs are produced and must be purified very quickly using generic methods disregarding analytical screening of various columns and accommodating lower recovery and purity. The introduction of the new SFC-MS Prep-100 system by Thar (now Waters) has now set a new standard for this kind of purification.

More recently, Rosse et al. successfully developed a platform to purify more than 90,000 samples (10–20 mg amounts) per year using this MS-based preparative SFC instrument and applying a generic method [148].

2.6 Theoretical considerations of preparative SFC

Given the particular physical properties of carbon dioxide in its supercritical state, especially high diffusity, it is expected that preparative SFC processes differ from classical preparative HPLC. In this context, a number of research groups became interested in studying the factors that might influence the performance, productivity, and robustness of preparative SFC applications. These theoretical considerations are particularly important for large-scale SFC, which necessitates stable processes delivering the desired quality and productivity in a continuous and unattended way. A better understanding of the chromatographic parameters also permits the elaboration of modeling tools to help optimize chromatographic processes. Moreover, theoretical knowledge is also very valuable for small-scale SFC purifications, considering that a poor performance might affect the efficiency of the purification process, leading to a lower thoughput.

This section very briefly presents some theoretical efforts performed in this field, with a focus on the preparative aspects.

In 2003, Rajendran, Mazzotti, and Morbidelli published a paper pointing to the critical factors affecting SFC on packed bed. They developed models to characterize adsorption equilibria, mass transfer, and column dynamics, and investigated the effect of polar modifiers [158]. Deviations with respect to HPLC and GC were discussed. The effect of pressure drop on solute retention and column efficiency was also studied by the same groups in Zurich [159–160].

Similarly, Brunner and Johannsen reported on new insights on adsorption and desorption processes in SFC [161]. They developed a model to describe adsorption equilibria.

Over the last five years, a number of systematic investigations were performed to facilitate the transfer from analytical-scale SFC to preparative SFC [162–164]. The influence of process parameters such as pressure, temperature of the fluid as well as type and concentration the modifier on SFC performance were examined. With their design of experiments approach, Enmark et al. concluded that the methanol fraction and pressure are critical parameters to be controlled to preserve retention throughout the scale-up [163]. The same research group developed a chemometric approach to study the combined effect of temperature, pressure, and co-solvent fraction in analytical and preparative SFC [164].

Guiochon and his group also investigated experimentally and by simulation the effect of pressure, temperature, and density drops along SFC columns [165–166]. In 2011, Guiochon and Tarafder reviewed the fundamentals of preparative SFC in a very exhaustive paper [167]. Recently, Fornstedt reviewed the status and subsisting challenges of preparative SFC [168], and in particular the multiple causes for peak distortions in preparative SFC [169].

With the continuous expansion of preparative SFC, it can reasonably be expected that the technology will soon be implemented in pilot and production and that such fundamental will be extremely helpful for process optimization.

2.7 Conclusion

Preparative packed SFC technology has been around for a long time but only sporadically applied in the previous millennium. Applications included small-scale purification of polymers, natural products, and drug impurities for characterization, but also very large applications such as the purification of DHA and EPA from fish oil. Since the beginning of the new millennium, the interest in preparative SFC for chiral separation continuously grew and it is now the technique of choice for this application from mg to kg scale. For achiral purifications of low and medium molecular weight molecules, the successful implementation of preparative MS-triggered SFC platforms in many laboratories across the world has now led to a trend reversal of the purification strategy. SFC has also evolved as the technique of first choice, whereas reversed phase chromatography (RP-HPLC) is the second choice. It can be estimated that hundreds of thousands of sample mixtures are currently purified yearly in the medicinal chemistry by means of achiral SFC. This revolution is probably remained invisible for many academic researchers considering that most of the applications are performed in the industry and have not been disclosed. For very large-scale SFC applications, there is still a need for exploring the scope and limitation of the technique, even though earlier large-scale applications have already demonstrated the feasibility. SFC is clearly changing the face of the purification world, and if it fits the purpose, SFC should be preferred considering the incontestable environmental and cost advantages. Anyway, SFC has now surmounted its hurdles and it is unlikely that SFC comes back in a depression phase as it was regularly the case since its first application in 1962. This is of course encouraging in the context of the growing awareness of our society and the large debate about the global warming and waste of energy that has triggered a large consensus regarding the promotion and support of sustainable processes consuming less energy.

References

[1] E. Klesper, A.H. Corwin, D.A. Turner, High pressure gas chromatography above critical temperatures, J. Org. Chem. 27 (1962) 700–701.

[2] S.T. Sie, W. van Beersum, G.W.A. Rijinders, High-pressure gas chromatography and chromatography with supercritical fluids. I. The effect of pressure on partition coefficients in gas-liquid chromatography with carbon cioxide as a carrier gas, Sep. Sci. 1 (1966) 459–490.

[3] J.C. Giddings, M.N. Myers, L. McLaren, R.A. Keller, High pressure gas chromatography of nonvolatile species, Science 162 (1968) 67–73.

[4] Wikipedia web page: https://en.wikipedia.org/wiki/Critical_point_(thermodynamics) (accessed 25 October 2017).

[5] M. Perrut, Procédé de fractionnement d'un mélange par chromatographie d'élution avec un fluide en état supercritique, Patent FR 8209649 (1982). US Pat. 4478720, October 1984.

[6] C.M. Harris, Product review: the SFC comeback, Anal. Chem. 74 (2002) 87A–91A.

[7] T.A. Berger, J.F. Deye, Separation of benzenepolycarboxylic acids by packed column supercritical fluid chromatography using methanol-carbon dioxide mixtures with very polar additives, J. Chromatogr. Sci. 29 (1991) 141–146.

[8] T.A. Berger, J.F. Deye, Effects of column and mobile phase polarity using steroids as probes in packed-column supercritical fluid chromatography, J. Chromatogr. Sci. 29 (1991) 280–286.

[9] T.A. Berger, J.F. Deye, Effect of basic additives on peak shapes of strong bases separated by packed-column supercritical fluid chromatography, J. Chromatogr. Sci. 29 (1991) 310–317.

[10] E. Francotte, Practical aspects and applications of preparative supercritical fluid chromatography, in: C.F. Poole (Ed.), Supercritical Fluid Chromatography, Handbooks in Separation Science, Elsevier Inc.: Amsterdam, 2017, pp. 275–316.

[11] D. Speybrouck, E. Lipka, Preparative supercritical fluid chromatography: a powerful tool for chiral separations, J. Chromatogr. A 1467 (2016) 33–55.

[12] E. Lemasson, S. Bertin, C. West, The use and practice of achiral and chiral supercritical fluid chromatography in pharmaceutical analysis and purifications, J. Sep. Sci. 39 (2016) 212–233.

[13] S. Borman, Acetonitrile shortage hurts research laboratories, Chem. Eng. News, 68 (1990) 15.

[14] R.E. Majors, The continuing acetonitrile shortage: how to combat it or live with it, LCGC North Am. Column Watch, 27 (2009) 450–471.

[15] C. Berger, M. Perrut, Preparative supercritical fluid chromatography, J. Chromatogr. 505 (1990) 37–43.

[16] L.T. Taylor, Supercritical fluid chromatography for the 21st century. J. Supercritical Fluids 47 (2009) 566–573.

[17] T.A. Berger, The past, present, and future of analytical supercritical fluid chromatography. Chromatogr. Today August/September 2014, 26–29.

[18] D.J. Tognarelli, An evaluation of physico-chemical properties of the mobile phase in supercritical fluid chromatography when using sub-2-mm particles columns, Jasco Application note, 2013. https://jascoinc.com/wp-content/uploads/2013/11/An-evaluation-of-physico-chemical-properties-of-the-mobile-phase-in-Supercritical-Fluid-Chromatography-when-using-sub-particle-columns.pdf.

[19] E. Brunner, Fluid mixtures at high pressures I. Phase separation and critical phenomena of 10 binary mixtures (a gas + methanol), J. Chem. Thermod. 17 (1985) 671–679.

[20] P. Mourier, E. Eliot, M. Caude, R. Rosset, A. Tambute, Supercritical and subcritical fluid chromatography on a chiral stationary phase for the resolution of phosphine oxide enantiomers, Anal. Chem. 57 (1985) 2819–23.

[21] A. Grand-Guillaume Perrenoud, C. Hamman, M. Goel, JL.Veuthey, D. Guillarme, S. Fekete, Maximizing kinetic performance in supercritical fluid chromatography using state-of-the-art instruments, J. Chromatogr. A 1314 (2013) 288–297.

[22] V. Desfontaine, D. Guillarme, E. Francotte, L. Nováková, Supercritical fluid chromatography in pharmaceutical analysis, J. Pharm. Biomed. Anal. 113 (2015) 56–71.

[23] E. Francotte, Practical advances in SFC for the purification of pharmaceutical molecules, LCGC Eur. 29 (2016) 194–204.

[24] T. Wang, M. Barber, I. Hardt, D.B. Kassel, Mass-directed fractionation and isolation of pharmaceutical compounds by packed-column supercritical fluid chromatography/mass spectrometry, Rap. Comm. Mass Spect. 15 (2001) 2067–2075.

[25] X. Zhang, M.H. Towle, C.E. Felice, J.H. Flament, W.K. Goetzinger, Development of a mass-directed preparative supercritical fluid chromatography purification system, J. Comb. Chem. 8 (2006) 705–714.

[26] R. Maiefski, D. Wendell, W.C. Ripka, J.D. Krakover, Apparatus and method for multiple channel high throughput purification, PCT Int. Appl. WO 00/266622, 2000.

[27] R. Chen, P. Ridgway, A case study of using Thar SFC-MS Prep 30® to purify polar, basic pharmaceutical relevant compounds, LC.GC North America, The Application Notebook, 2008, 01 September. http://www.chromatographyonline.com/case-study-using-thar-sfc-ms-prep-30-purify-polar-basic-pharmaceutical-relevant-compounds.

[28] L. Miller, I. Sebastian, Evaluation of injection conditions for preparative supercritical fluid chromatography, J. Chromatogr. A 1250 (2012) 256–263.

[29] M. Shaimi, G.B. Cox, Mixed stream vs modifier stream injections. PIC Solution Inc., http://www.pic-sfc.com/publications1.html (accessed 25 October 2017), Application Note 6/13.

[30] Y. Horikawa Y, K. Kamezawa, T. Kanomata, M. Bounoshita, M. Saito, Novel injection method in preparative supercritical fluid chromatography. Poster presented at the 2nd International Conference on Packed-Column SFC. Zurich, Switzerland, 1–2 October 2008. Available from http://www.jascoinc.com/docs/application-notes/InjectionMethod_PrepSFC.pdf.

[31] M. Shaimi, G.B. Cox, Injection by extraction: a novel sample introduction technique for preparative SFC, Chromatogr. Today November/ December 2014, Vol. 7, 42–45.

[32] G.B. Cox, Injection by extraction in preparative SFC; isolation of impurities and natural products. Chromatogr. Today August/ September 2016, Vol. 9, 16–21.

[33] Y. Dai, G. Li, A. Rajendran, Peak distortions arising from large-volume injections in supercritical fluid chromatography, J. Chromatogr. A 1392 (2015) 91–99.

[34] Y. Hirata, Y. Kawaguchi, K. Kitano, Large volume injection for preparative supercritical fluid chromatography, Chromatographia 40 (1995) 42–46.

[35] M. Perrut, Procédé d'extraction-séparation-fractionnement par fluides supercritiques et dispositif pour sa mise en oeuvre. French Patent 8,510,468, (1985); Eur. Pat. 8,640,139, (1986).

[36] A. Bozic, Conversion of a Standard Preparative HPLC into a Preparative SFC. Lecture at the 6th International Conference on Packed-Column SFC. Brussels, Belgium, 3–5 October 2012. http://www.greenchemistrygroup.org/past-conferences.

[37] R. Chen R, J. Cole, Feasibility of using ELSD to trigger fraction collection in small-scale purification by SFC. LC-GC Eur., The Application Notebook July/August 22 (2009) 33.

[38] J. Thompson, B. Strode III, L.T. Taylor, K. Anton, M. Bach, N. Pericles, Packed columns supercritical fluid chromatography-evaporative light scattering detection, in: K. Anton, C. Berger (Eds.), Supercritical Fluid Chromatography with Packed Columns. Chromatographic Science Series, Marcel Dekker, New York, 1997, Vol. 75, pp. 97–123.

[39] G.W. Yanik, Polarimetric detection in supercritical fluid chromatography. in: Webster GK, editor. Advances and Applications in Pharmaceutical Analysis: Supercritical Fluid Chromatography, Pan Standford Publishing:Singapore, 2014, pp. 333–346.

[40] E.S. Wentz, Multimode collections in prep SFC working towards open access SFC Sub-2 um particles for SFC. First International Conference on Packed Column SFC, Pittsburgh, 23–25 September 2007. Slides available from http://www.greenchemistrygroup.org/past-conferences.

[41] T. Kanomata, C. Silverman, Y. Horikawa, H. Saito, M. Bounoshita, M. Saito, Advantages of circular dichroism (CD) detection for the determination of fractionation timing in preparative supercritical fluid chromatography (Prep-SFC) for chiral separations. Poster presented at the First International Conference on Packed Column SFC, Pittsburgh, 23–25 September 2007. Slides available from http://www.greenchemistrygroup.org/past-conferences or LCGC The Peaks, December (2007) 7–22.

[42] Y. Hirata, Y. Kawagushi, Y. Funada, Refractive index detection using and ultraviolet detector with a capillary flow cell in preparative SFC, J. Chromatogr. Sci. 34 (1996) 58–62.

[43] C.F. Poole, Stationary phases for packed-column supercritical fluid chromatography, J. Chromatogr. A 1250 (2012) 157–171.

[44] C. West, E. Lesellier, Characterisation of stationary phases in subcritical fluid chromatography with the solvation parameter model IV: Aromatic stationary phases, J. Chromatogr. A 1115 (2006) 233–245.

[45] C. West, E. Lesellier, A unified classification of stationary phases for packed column supercritical fluid chromatography, J. Chromatogr. A 1191 (2008) 21–39.

[46] C. West, E. Lesellier, Chemometric methods to classify stationary phases for achiral packed column supercritical fluid chromatography, J. Chemom. 26 (2012) 52–65.

[47] S. Khater, C. West, E. Lesellier, Characterization of five chemistries and three particle sizes of stationary phases used in supercritical fluid chromatography, J. Chromatogr. A 1319 (2013) 148–159.

[48] C. West, M.A. Khalikova, E. Lesellier, K. Heberger, Sum of ranking differences to rank stationary phases used in packed column supercritical fluid chromatography, J. Chromatogr. A 1409 (2015) 241–250.

[49] E. Francotte, Why should SFC be the first technology choice for the purification of pharmaceuticals?, Lecture at the 8th International Conference on Packed-Column SFC, Basel, Switzerland, 8–10. October. 2014. Slides available from http://media.wix.com/ugd/2239fc_6eee04127c92436ab62c550d8b8a9317.pdf

[50] L. Miller, Chem Oggi, Pharmaceutical purifications using preparative supercritical fluid chromatography, 32(2) (2014) 23–26.

[51] E. Francotte, G. Diehl, SFC: The universal purification technique? lecture at the 4th international conference on packed-column SFC, Stockholm, Sweden September 15–16, (2010). Slide available can be loaded from http://media.wix.com/ugd/2239fc_fc3a01aa76d5458a86c0f0eaeac983e8.pdf.

[52] E.L. Regalado, C.J. Welch. Separation of achiral analytes using supercritical fluid chromatography with chiral stationary, Trends Anal. Chem. 67 (2015) 74–81.

[53] T. Shibata, S. Shiruka, A. Ohnishi, Y. Mukarami, K. Ueda, Achiral SFC on polysaccharide phases. Lecture presented at the 9th International Conference on Packed column SFC, 23 July . 2015, Philadelphia (USA). Slides available from http://media.wix.com/ugd/2239fc_b5a4641c46d24e92a3b2e6fdd3d6c6d1.pdf.

[54] T. Shibata, S. Shinkura, A. Ohnishi, K. Ueda, Achiral molecular recognition of aromatic Position isomers by polysaccharide-based CSPs in relation to chiral recognition, Molecules 22 (2017) 38.

[55] L. Leith, D-R. Wu, P. Li, D. Sun, Benzoate SFC separations on chiral and achiral stationary phases, Poster presented at the 3rd International Conference on Packed-Column SFC, 22–23. July 2009, Philadelphia (USA). http://www.greenchemistrygroup.org/past-conferences.

[56] M. Biba, J. Liu, A Perspective on the application of preparative supercritical fluid chromatography using achiral stationary phases in pharmaceutical drug discovery and development, Am. Pharm. Rev. 19 (3) April (2016).

[57] W. Hartmann, E. Klesper, Preparative supercritical fluid chromatography of styrene oligomers, J. Polymer Sci., Polymer Lett. Ed. 15 (1977) 713–719.

[58] P. Jusforgues, C. Berger, M. Perrut, New separation process: preparative supercritical fluid chromatography, Chem. Ingen. Techn. 59 (1987) 666–667.

[59] P. Jusforgues, Preparative supercritical fluid chromatography: principles, potential, and development, Spectra 2000 166 (1992) 53–56.

[60] M. Saito, Y. Yamauchi, Isolation of tocopherols from wheat germ oil by recycle semi-preparative supercritical fluid chromatography, J. Chromatogr. 505 (1990) 257–271.

[61] Y. Yamauchi, M. Saito, Fractionation of lemon-peel oil by semi-preparative supercritical fluid chromatography, J. Chromatogr. 505 (1990) 237–246.

[62] K. Ute, N. Miyatake, T. Asada, K. Hatada, Stereoregular oligomers of methyl methacrylate. 6. Isolation of isotactic and syndiotactic methyl methacrylate oligomers from 19–mer to 29–mer by preparative supercritical fluid chromatography and their thermal analysis, Polym. Bull. (Berlin, Germany) 28 (1992) 561–568.

[63] G. Fritz, G. Feucht, Analytical and preparative separation of functional carbosilanes and phosphines by means of SFC (supercritical fluid chromatography), Zeit. Anorg. Allgem. Chem. 593 (1991) 69–89.

[64] M. Hanson, Small-scale preparative supercritical fluid chromatography of cyproterone acetate, Chromatographia 40 (1995) 139–142.

[65] M. Hanson, Micropreparative supercritical fluid chromatography as a bench-scale method for steroid purification, LC-GC 14 (1996) 152–158.

[66] E. Ihara, M. Tanabe, Y. Nakayama, A. Nakamura, H. Yasuda, Characterization of lactone oligomers isolated by preparative SFC, Macrom. Chem. Phys. 200 (1999) 758–762.

[67] I. Flament, U. Keller, L. Wunsche, Use of semi-preparative supercritical fluid chromatography for the separation and isolation of flavor and food constituents, in: S.H. Rizvi (Ed.), Supercritical Fluid Processing of Food and Biomaterials, 1994, pp. 62–74.

[68] G. Cretier, J. Neffati, J.L. Rocca, Preparative LC and preparative SFC: two complementary techniques in the fractionation of an impurity from a major component, J. Chromatogr. Sci. 32 (1994) 449–454.

[69] G.B. Jacobson, K.E. Markides, B. Langstrom, Supercritical fluid synthesis and online preparative supercritical fluid chromatography of 11C-labeled compounds in supercritical ammonia, Acta Chem. Scand. 51 (3, Suppl.) (1997) 418–425.

[70] Y. Kadota, I. Tanaka, Y. Ohtsu, M. Yamaguchi, Separation of polyunsaturated fatty acids by chromatography using a silver-loaded spherical clay. I. Pilot-scale preparation of high-purity docosahexaenoic acid by supercritical fluid chromatography, Nihon Yukagakkaishi 46 (1997) 397–403

[71] Y. Wang, G. Stutz, S. Roggo, G. Diehl, E. Francotte, Development of SFC for high throughput purification in natural product drug discovery. Lecture presented at the 2nd International Conference on Packed-Column SFC, Zurich, Switzerland, 1–2 October 2008. Slides available from http://media.wix.com/ugd/2239fc_c1832bef0ef548aba914f87bc86c870d.pdf

[72] W. Song, X. Qiao, W.F. Liang, S. Ji, L. Yang, Y. Wang, Y. Xu, Y. Yang, D. Guo, M. Ye, Efficient separation of curcumin, demethoxycurcumin, and bisdemethoxycurcumin from turmeric using supercritical fluid chromatography: from analytical to preparative scale, J. Sep. Sci. 38 (2015) 3450–3453.

[73] G. Vicente, M.R. Garcia-Risco, T. Fornari, G. Reglero, Isolation of carsonic acid from rosemary extracts using semi-preparative supercritical fluid chromatography, J. Chromatogr. A 1286 (2013) 208–215.

[74] O.I. Pokrovskii, A.A. Krutikova, K.B. Ustinovich, O.O. Parenago, M.V. Moshnin, S.A. Gonchukov, V.V. Lunin, Preparative separation of methoxy derivatives of psoralen using supercritical-fluid chromatography, Rus. J. Phys. Chem. B 7 (2013) 901–915.

[75] S. Li, T. Lambros, Z. Wang, R. Goodnow, C.-T. Ho, Efficient and scalable method in isolation of polymethoxyflavones from orange peel extract by supercritical fluid chromatography, J. Chromatogr. B 846 (2007) 291–297.

[76] Z. Wang, Supercritical Fluid Chromatography – A powerful tool for polymethoxyflavone analysis and isolation. Lecture presented at the 3rd International Conference on Packed-Column SFC. Philadelphia, PA, USA. July 22–23, 2009. Slides available from http://media.wix.com/ugd/2239 fc_180972ec65134882a1acad3b88b03557.pdf

[77] M.A. Patel, F. Riley, M. Ashraf-Khorassani, L.T. Taylor, Supercritical fluid chromatographic resolution of water soluble isomeric carboxyl/amine terminated peptides facilitated via mobile phase water and ion pair formation, J. Chromatogr A 1233 (2012) 85–90.

[78] F. Montanes, P. Rose, S. Tallon, R. Shirazi, Separation of derivatized glucoide anomers using supercritical fluid chromatography, J. Chromatogr. A 1418 (2015) 218–223.

[79] G. Terfloth, Preparative isolation of impurities, in: R.J. Smith, M.L. Webb (Eds), Analysis of Drug Impurities, Blackwell Publishing, Oxford, UK, 2007, pp 215–234.

[80] T. Zelesky, Supercritical fluid chromatography (SFC) as an isolation tool for the identification of drug related impurities, Am. Pharm. Rev. 11 (2008) 56–60, 62.

[81] S. Buskov, J. Hasselstrom, C.E. Olsen, H. Sorensen, J.C. Sorensen, S. Sorensen, Supercritical fluid chromatography as a method of analysis for the determination of 4–hydroxybenzylglucosinolate degradation products, J. Biochem. Biophys. Meth. 43 (2000) 157–174

[82] S. Klobcar, H. Prosen, Isolation of oxidative degradation products of atorvastatin with supercritical fluid chromatography, Biomed. Chromatogr. 29 (2015) 1901–1906.

[83] K. Shimada, M.A. Lusenkova, K. Sato, T. Saito, S. Matsuyama, H. Nakahara, S. Kinugasa, Evaluation of mass discrimination effects in the quantitative analysis of polydisperse polymers by matrix-assisted laser desorption/ionization time-of-flight mass spectrometry using uniform oligostyrenes, Rapid Commun. Mass Spectrom. 15 (2001) 277–282.

[84] W. Eckni, H.J. Polster, Supercritical chromatography of paraffins on a molecular sieve: analytical and preparative scale, Sep. & Sci. Techn. 21 (1986) 139–156.

[85] K. Jinno, H. Nagashima, K. Itoh, M. Saito, M. Buonoshita, Subcritical fluid extraction and supercritical fluid chromatography of carbon clusters C60 and C70, Fresenius' Journal of Analytical Chemistry 344 (1992) 435–441.

[86] K. Ute, K. Hatada, Applications of preparative SFC to oligomer analysis and characterization in: M. Saito, Y. Yamauchi, T. Okuyama (Eds.), Fractionation Packed-Column SFC SFE, VCH:New York, 1994, pp 231–253.

[87] K. Hatada, T. Kitayama, K. Ute, K. Nishiura. Uniform polyolefin and polymethacrylate, Macromol. Symp. 129 (1998) 89–97.

[88] T. Kitayama, K. Nishiura, M. Tsubota, H. Nakanishi, K. Hatada. Uniform regular and irregular three-arm star poly(methyl methacrylate)s, Polymer J. 31 (1999) 1001–1004.

[89] K. Hatada, T. Kitayama. Structurally controlled polymerizations of methacrylates and acrylates, Polym. Int. 49 (2000) 11–47

[90] Y. Sasanuma, T. Iwata, H. Kato,T. Yarita,S. Kinugasa, R.V. Law, carbon-13 NMR chemical shifts of dimeric model compounds of poly(propylene Oxide): A proof of existence of the (C–H)···O attraction, J. Phys. Chem. A 105 (2001) 3277–3283.

[91] K. Shimada, R. Nagahata, S. Kawabata, S. Matsuyama, T. Saito, S. Kinugasa, *Evaluation of the quantitativeness of matrix-assisted laser desorption/ionization time-of-flight mass spectrometry using an equimolar mixture of uniform poly(ethylene glycol) oligomers,* J. Mass Spectrom. 38 (2003) 948–954.

[92] E. Francotte, SCF: A many-sided technology to support drug discovery, Lecture at the 10th International conference on packed-column SFC, Vienna (Austria), 5–17 October 2016. http://www.greenchemistrygroup.org/past-conferences.

[93] E. Briard, D. Orain, C. Beerli, A. Billich, M. Streiff, M. Bigaud, Y.P. Auberson, BZM055, an iodinated radiotracer candidate for PET and SPECT imaging of myelin and FTY720 brain distribution, ChemMedChem 6 (2011) 667–677.

[94] M. Ashraf-Khorassani, M.B. Gandee, M.T. Combs, L.T. Taylor, Stationary phases for packed-column supercritical fluid chromatography, J. Chromatogr. Sci. 35 (1997) 593–596.

[95] J.O. DaSilva, H.S. Yip, V. Hegde, A. Zaks, Supercritical fluid chromatography (SFC) as a green chromatographic technique for support in rapid development of pharmaceutical candidates, Lecture presented at the 2nd International Conference on Packed-Column SFC, Zurich (Switzerland), 1–2 October 2008. http://www.greenchemistrygroup.org/past-conferences.

[96] S. Higashidate, Y. Yamauchi, M. Saito, Enrichment of eicosapentaenoic acid and docosahexaenoic acid esters from esterified fish oil by programmed extraction-elution with supercritical carbon dioxide, J. Chromatogr. 515 (1990) 295–303.

[97] Y. Kadota, I. Tanaka, Y. Ohtsu, M. Yamaguchi, Separation of polyunsaturated fatty acids by chromatography using a silver-loaded spherical clay. I. Pilot-scale preparation of high-purity docosahexaenoic acid by supercritical fluid chromatography. Nihon Yukagakkaishi 46 (1997) 397–403.

[98] M. Alkio, C. Gonzalez, M. Jantti, O. Aaltonen, Purification of polyunsaturated fatty acid esters from tuna oil with supercritical fluid chromatography, J. Am. Oil Chem. Soc. 77 (2000) 315–321.

[99] G. Pettinello, A. Bertucco, P. Pallado, A. Stassi, Production of EPA enriched mixtures by supercritical fluid chromatography: from the laboratory scale to the pilot plant, J. Supercritical Fluids 19 (2000) 51–60.

[100] R. Krumbholz, P. Lembke, N. Schirra, Separation of fatty acids using preparative supercritical fluid chromatography. PCT Int. Appl. (2007), WO 2007/147554 A2.

[101] P. Lemke, Production of high purity n-3 fatty acid-ethyl esters by process scale supercritical fluid chromatography, in: K. Anton, C. Berger (Eds.), Supercritical Fluid Chromatography with Packed Columns. Chromatographic Science Series, Marcel Dekker:New York, 1997, Vol 75, pp. 429–443.

[102] F. Montanes, O.J. Catchpole, S. Tallon, K. Mitchell, K. Lagutin, Semi-preparative supercritical chromatography scale plant for polyunsaturated fatty acids purification, J. Supercritical Fluids 79 (2013) 46–54.

[103] M. Saito, Y. Yamauchi, K. Inomata, W. Kottkamp, Enrichment of tocopherols in wheat germ by directly coupled supercritical fluid extraction with semipreparative supercritical fluid chromatography, J. Chromatogr. Sci. 27 (1989) 79–85.

[104] P. Jusforgues, M. Shaimi, D. Barth, Preparative supercritical fluid chromatography: grams, kilograms, and tons!, in: K. Anton, C. Berger (Eds.), Supercritical Fluid Chromatography with Packed Columns. Chromatographic Science Series, Marcel Dekker:New York, 1997, Vol 75, pp. 414–415.

[105] A. Depta, T. Giese, M. Johannsen, G. Brunner, Separation of tereoisomers in a imulated oving ed-upercritical luid hromatography lant, J. Chromatogr. A 865 (1999) 175–186.

[106] M. Ashraf-Khorassani, L.T. Taylor, M. Martin, Supercritical fluid extraction of ava lactones from kava root and their separation via supercritical fluid chromatography, Chromatographia 50 (1999) 287–292.

[107] M. Johannsen, Process for the preparation of vitamin D3 and provitamin D3, Eur. Pat. Appl. (2000), EP 969001 A2 20000105.

[108] S.L. Taylor, J.W. King, Preparative-scale supercritical fluid extraction/supercritical fluid chromatography of corn bran, J. Am. Oil Chem. Soc. 79 (2002) 1133–1136.

[109] P. Ramirez, M.R. Garcia-Risco, S. Santoyo, F.J. Senorans, E. Ibanez, G. Reglero, Isolation of functional ingredients from rosemary by preparative-supercritical fluid chromatography (Prep-SFC), J. Pharm. Biom. Anal. 41 (2006) 1606–1613.

[110] P. Ramirez, S. Santoyo, M.R.Garcia-Risco, F.J. Senorans, E. Ibanez, G. Reglero, Use of specially designed columns for antioxidants and antimicrobials enrichment by preparative supercritical fluid chromatography, J. Chromatogr. A (2007) 1143 234–242.

[111] G. Vicente, M.R. Garcia-Risco, T. Fornari, G. Reglero, Isolation of carsonic acid from rosemary extracts using semi-preparative supercritical fluid chromatography, J. Chromatogr. A 1286 (2013) 208–215.

[112] M.H. Ng, Y.M. Choo, H.Yahaya, Pilot scale supercritical fluid chromatography: an experience. Lecture presented at the 2nd International Conference on Packed Column SFC. Zurich (Switzerland), 1–2 October 2008. Slides available from http://media.wix.com/ugd/2239fc_8387fad969ff45afab2a826ab7712a4a.pdf.

[113] M. Ashraf-Khorassani, N. Nazem, L.T. Taylor, W.M. Coleman, Isolation, fractionation, and identification of sucrose esters from various oriental tobaccos employing supercritical fluids,Beitraege zur Tabakforschung International 23 (2008) 32–45.

[114] H. Lu, X. Mo, M. Zhang, M. Yang, Purification of citric acid by supercritical fluid chromatography, Shipin Gongye Keji 33 (2012) 271–274.

[115] L. Wunsche, U. Keller, I. Flament, Combination of supercritical fluid chromatography with thin-layer chromatography on a semi-preparative scale, J. Chromatogr. A 552 (1991) 539–549.

[116] F. Pang, H. Lü, Y. Liu, M. Zhang, The study on purification of artemisinin by supercritical fluid chromatography, J. Chem. Eng. Chin. Univ. 24 (2010) 569–573.

[117] M.T. Liang, R.C. Liang, S. Yu, R. Yan, Separation of resveratrol and emodin by supercritical fluid-simulated moving bed chromatography, J. Chromat Separation Techniq 4 (2013) 1–5.

[118] J. Lundell, J. Ruohonen, O. Aaltonen, M. Alkio, A. Hase, T. Suortti, Process for preparation of pure cyclosporin chromatographically using an eluent consisting essentially of high pressure carbon dioxide. WO 1996027607 A1, Priority date, Mach 3. 1995, publ. 12 September 1996.

[119] O. Aaltonen, M. Alkio, J. Lundell, J. Ruohonen, L. Parvinen, V. Suoninen, Polypeptide purification with industrial-scale supercritical fluid chromatography. Pharm. Techn. Europe, 10 (1998) A42–A54.

[120] P. Jusforgues, M. Shaimi, D. Barth, Preparative supercritical fluid chromatography: grams, kilograms, and tons! in: K. Anton, C. Berger (Eds.), Supercritical Fluid Chromatography with Packed Columns. Chromatographic Science Series, Marcel Dekker:New York, 1997, Vol 75, pp 410–414.

[121] T. Imahashi, Y. Yamauchi, M. Saito, Fractionation and identification of additives in poly(vinyl chloride) film by supercritical fluid extraction/semi-preparative supercritical fluid chromatography, Bunseki Kagaku 39 (1990) 79–85.

[122] L. Miller, Flash chromatography, Lecture at the 4th International Conference on SFC, 2010, Stockholm (Sweden), 15–16 September 2010. http://www.greenchemistrygroup.org/past-conferences.

[123] L. Miller, M. Mahoney, Evaluation of flash supercritical fluid chromatography and alternate sample loading techniques for pharmaceutical medicinal chemistry purifications, J. Chromatogr. A 1250 (2012) 264–73.

[124] M. Ashraf-Khorassani, Q. Yan, A. Akin, F. Riley, C. Aurigemma, L.T. Taylor, Feasibility of correlating separation of ternary mixtures of neutral analytes via thin layer chromatography with supercritical fluid chromatography in support of green flash separations, J. Chromatogr. A 1418 (2015) 210–217.

[125] F. Riley, T. Zelesky, B. Marquez, C. Brunelli, Packed column SFC: impacting project progression from drug discovery through development. Lecture presented at the 2nd International

conference on packed column SFC. Zurich (Switzerland), 1–2 October 2008. http://www.greenchemistrygroup.org/past-conferences.

[126] M. Przybyciel, J. Stefkovich, R. Schlake, Development and application of a novel 'green' flash chromatography purification system, ES Industries (esind.com), Xenosep Technologies & Applied Separations, Poster presentation at the 8th International Conference on Packed-Column SFC, Basel, Switzerland, 8–10. October. 2014. http://www.greenchemistrygroup.org/past-conferences.

[127] R. McClain, V. Rada, A. Nomland, M. Przybyciel, D. Kohler, R. Schlake, P. Nantermet, C.J. Welch, Greening Flash Chromatography, ACS Sustainable Chem. Eng., 4 (2016) 4905–4912.

[128] M.A. Bums, Green Flash SFC. Poster presented at the 3rd International Conference on Packed-Column SFC, 22–23.07.2009, Philadelphia, USA. http://www.greenchemistrygroup.org/past-conferences.

[129] M.H. Ng, Y.M. Choo, Chromatography for the analyses and preparative separations of palm oil minor components, Am. J. Anal. Chem. 6 (2015) 645–650.

[130] J.Y. Clavier, R.M. Nicoud, M. Perrut, A new efficient fractionation process: the simulated moving bed with supercritical eluent, in: P.R. Rohr, C. Trepp (Eds.), High Pressure Chemical Engineering, Elsevier Science, London, 1996, pp 429–434.

[131] E. Valery, W. Majeswski, C. Morey, H. Osuna-Sanchez, Mi. Bailly, SupersepTM Max: Breakthrough multi-column SFC technology, presented at the 2nd International Conference on Packed Column SFC. Zurich (Switzerland), 1–2 October 2008. http://www.greenchemistrygroup.org/past-conferences.

[132] S. Peper, M. Johannsen, G. Brunner, Preparative chromatography with supercritical fluids. Comparison of simulated moving bed and batch processes, J. Chromatogr. A 1176 (2007) 246–253.

[133] M. Johanssen, Supercritical fluid chromatography: from analytical tool to industrial separation processes. Lecture presented the 4th International Conference on Packed-Column SFC, September 15–16, 2010 Stockholm, Sweden. Slides available from http://media.wix.com/ugd/2239fc_ca204fa5d0e146128a97c132d290d00f.pdf

[134] M.T. Liang, R.C. Liang, L.R. Huang, P.H. Hsu, Y.H. Wu, H.E. Yen, Separation of sesamin and sesamolin by a supercritical fluid-simulated moving bed, Am. J. Anal. Chem. 3 (2012) 931–938.

[135] P. Searle, K.A.Glass, J. Hochlowski, Comparison of preparative HPLC/MS and preparative SFC techniques for the high-throughput purification of compound libraries, J. Comb. Chem. 6 (2004) 175–180.

[136] C. White, J. Burnett, Integration of supercritical fluid chromatography into drug discovery as a routine support, J. Chromatog. A 1074 (2005) 175–185.

[137] C. Aurigemma, Breaking the flow rate barrier: utilization of high flow mass-directed SFC for pharmaceutical applications, Lecture at the 3rd International Conference on Packed-Column SFC. Philadelphia, PA, USA. July 22–23, 2009. Slides available from http://media.wix.com/ugd/2239fc_af89841a5cc843f4842e24142ce3bec0.pdf

[138] R. McClain, J. Small, several uses of mass-directed SFC in basic research at merck, Paper presented at the 3rd International Conference Packed-Column SFC, 22–23 July 2009.

[139] L. Ma, V. Lazarescu, M.J. Mulvihill, Toward a "universal approach" for mass-directed SFC purification of small molecule compound libraries, Lecture at the 4th International Conference on Packed-Column SFC, 15–16 September 2010 Stockholm, Sweden. Slides available from http://media.wix.com/ugd/2239fc_764d60b4afcf4f73aec61e83f4918955.pdf

[140] J. Van Anda, SFC purification of achiral compounds – Pre-screening ro scale up. Lecture at the 4th International Conference on Packed-Column SFC, Stockholm, Sweden September 15–16, 2010. http://www.greenchemistrygroup.org/past-conferences.

[141] J. Van Anda, Use of SFC/MS in the purification of achiral pharmaceutical compounds, Am. Pharm. Rev. 13 (2010) 111–115.

[142] A. Mich, B. Matthes, R. Chen, S. Buehler, A comparative study on the purification of library compounds in drug discovery using mass-directed preparative SFC and preparative RPLC, LC-GC Eur. 2010, Vol. 23, 12–13.

[143] K Ebinger, HN Weller, J Kiplinger, P Lefebvre, Evaluation of a new preparative supercritical fluid chromatography system for compound library purification: the TharSFC SFC-MS Prep-100 system, J lab. automat. 16 (2011) 241–249.

[144] L. Miller, G. Bi, W. Goetzinger, Evaluation of non-traditional co-solvents for routine pharma discovery support. Paper presented at the 3rd International Conference Packed-Column SFC, 22–23 July 2009.

[145] D. Speybrouck, Application of SFC to the Analysis and Purification of Pharmaceutical Compounds. Lecture at the 4th International Conference on Packed-Column SFC, Stockholm, Sweden, 15–16 September (2010). http://www.greenchemistrygroup.org/past-conferences.

[146] Q.P. Han, M.J. Hayward, A highly automated 5 pump, 4 detector supercritical fluid chromatography (SFC) mas spectroscopy (MS) system for purification in drug discovery, Lecture at the 3rd International Conference on Packed-Column SFC, Poster presented at the 3rd International Conference Packed-Column SFC, 22–23. July 2009.

[147] E. Francotte, I. Adam, T. Mann, T Wolf, SFC Evolving as the Prevailing Purification Technique in Medicinal Chemistry. Lecture at the 6th International Conference on Packed-Column SFC, Brussels, Belgium, 4–5 October (2012). Slides available from http://greenchemistrygroup.org/pdf/2012/Francotte.pdf.

[148] G. Rosse, Maximizing efficiency in drug discovery: SFC-MS as the technique of choice for small molecules purification. Lecture at the 9th International Conference on Packed-Column SFC. Philadelphia, PA, USA, 23 July (2015). Slide available from http://media.wix.com/ugd/2239fc_9c99ce71471742199f2637d360a10cb5.pdf.

[149] M.L. de la Puente, P. López Soto-Yarritu, J. Burnett, Supercritical fluid chromatography in research laboratories: Design, development and implementation of an efficient generic screening for exploiting this technique in the achiral environment, J. Chromatogr. A 1218 (2011) 8551–8560.

[150] P. Korsgren P, Lanborg, A. Weinmann, Column selection for achiral purification using SFC-MS, Am. Pharm. Rev. 15 (2012) (4).

[151] R. McClain, M. Ho Hyun, Y. Li, C.J. Welch. Design, synthesis and evaluation of stationary phases for improved achiral supercritical fluid chromatography separations, J. Chromatogr. A 1302 (2013) 163–173.

[152] K. Ebinger, H.N. Weller, Comparative assessment of achiral stationary phases for high throughput analysis in supercritical fluid chromatography, J. Chromatogr. A 1332 (2014) 73–81.

[153] K. Coleman. High-throughput preparative separations from combinatorial libraries. Analusis 27 (1999) 719–723.

[154] T.A. Berger, K. Fogleman, T. Staats, P. Bente, I. Crocket, W. Farrell, M. Osonubi, The development of a semi-preparatory scale supercritical-fluid chromatograph for high-throughput purification of 'combi-chem' libraries, J Biochem. Biophys. Meth. 43 (2000) 87–111.

[155] J. Hochlowski, J. Olson, J. Pan, D. Sauer, P. Searle, T. Sowin, Purification of HTOS libraries by supercritical fluid chromatography, J Liq. Chromatogr. Rel. Techn. 26 (2003) 333–354.

[156] J. Hochlowski, High-throughput purification: Triage and optimization, in: B. Yan (Ed), Chemical Analysis John Wiley & Sons, (New York, NY, United States) 2004, 163 (Analysis and Purification Methods in Combinatorial Chemistry), pp. 281–306.

[157] M. Ventura, W. Farrell, C. Aurigemma, K. Tivel, M. Greig, J. Wheatley, A. Yanovsky, K.E. Milgram, D. Dalesandro, R. DeGuzman, High-throughput preparative process utilizing three complementary chromatographic purification technologies, J. Chromatogr. A 1036 (2004) 7–13.

[158] A. Rajendran, M. Mazzotti, M. Morbidelli, Preparative chromatography at supercritical conditions in: C.-H. Lee (Ed), Adsorption Science and Technology, Proceedings of the Pacific Basin Conference, 3rd, Kyongju, Republic of Korea, World Scientific, Singapore, May 25–29, 2003 (2003), pp. 204–208.

[159] A. Rajendran, O. Kräuchi, M. Mazzotti, M. Morbidelli, Effect of pressure drop on solute retention and column efficiency in supercritical fluid chromatography, J. Chromatogr. A 1092 (2005) 149–160.

[160] A. Rajendran, T.S. Gilkinson, M. Mazzotti, Effect of pressure drop on solute retention and column efficiency in supercritical fluid chromatography. Part 2: Modified carbon dioxide as mobile phase, J. Sep. Sci. 31 (2008) 1279–1289.

[161] G. Brunner, M. Johannsen, New aspects on adsorption from supercritical fluid phases, J. Supercritical Fluids 38 (2006) 181–200.

[162] M. Oman, P. Kotnik, M. Skerget, Z. Knez, Supercritical fluid chromatography and scale up study, Acta Chimica Slovenica 61 (2014) 746–758.

[163] M. Enmark, D. Aasberg, H. Leek, K. Oehlen, M. Klarqvist, J. Samuelsson, T. Fornstedt, Evaluation of scale-up from analytical to preparative supercritical fluid chromatography, J. Chromatogr. A 1425 (2015) 280–286.

[164] D. Aasberg, M. Enmark, J. Samuelsson, T. Fornstedt, Evaluation of co-solvent fraction, pressure and temperature effects in analytical and preparative supercritical fluid chromatography, J. Chromatogr. A 1374 (2014) 254–260.

[165] D.P. Poe, D. Veit, M. Ranger, K. Kaczmarski, A. Tarafder, G. Guiochon, Pressure, temperature and density drops along supercritical fluid chromatography columns. I. Experimental results for neat carbon dioxide and columns packed with 3- and 5-micron particles, J. Chromatogr. A 1250 (2012) 105–114.

[166] K. Kaczmarski, D.P. Poe, A. Tarafder, G. Guiochon, Pressure, temperature and density drops along supercritical fluid chromatography columns. II. Theoretical simulation for neat carbon dioxide and columns packed with 3-μm particles, J. Chromatogr. A 1250 (2012) 115–123.

[167] G. Guiochon, A. Tarafder. Fundamental challenges and opportunities for preparative supercritical fluid chromatography, J. Chromatogr. A 1218 (2011) 1037–1114.

[168] T. Fornstedt, Modern supercritical fluid chromatography – possibilities and pitfalls, LCGC North America 33, 1 (March 2015) 166–174.

[169] T. Fornstedt, Peak distortions in preparative supercritical fluid chromatography – a more complete overview, Chromatogr. Today, August/September 9 (2016) 10–14.

Abhijit Tarafder, Steven M. Collier, Jason F. Hill

3 Addressing robust operation and reproducibility in SFC

Abstract: One of the major factors that restricted supercritical fluid chromatography (SFC) being included as a part of regular analytical workflow is the lack of robust and reproducible operations. Modern packed-column SFC systems have the same structure as an ultrahigh-performance liquid chromatography system with the exception of deploying a pump specific for liquid or compressed CO_2 and an automated back-pressure regulator that keeps the entire system above a specified pressure. The main challenge that impeded robust SFC instrumentation for years is the task of handling liquid/compressed CO_2 which has significantly higher compressibility compared to liquid solvents. Although several aspects are often discussed under the topic of SFC instrumental challenges, they are different manifestations of the same factor – higher compressibility of CO_2. In addition to this, effects of chemical reactions in the presence of CO_2 and absence of water in imparting variability of stationary-phase surface properties have been highlighted [Fairchild et al. Chromatographic evidence of silyl ether formation (SEF) in supercritical fluid chromatography. Anal. Chem. 87 (2015) 1735–1742] recently. In this chapter, a brief overview on these issues and how they were addressed will be presented with special emphasis on the effect of compressibility on system performance variation and effect of chemical reactions on surface chemistry leading to retention drift in SFC.

Keywords: Compressibility, Isopycnic plot, Column packing, Torus technologies, Torus 2-PIC, Torus DEA, Torus DIOL, Torus 1-AA, Silyl-ether formation, robustness, DOE, QbD

3.1 Introduction

The main difference between an ultrahigh-performance liquid chromatography ((U) HPLC) system and that of a modern supercritical fluid chromatography (SFC) system is the use of CO_2 as the principal solvent in the mobile phase by the latter. Typically, one of the pumps in SFC is designed to handle liquid CO_2 while the other(s) are standard (U)HPLC pumps. Downstream of the pump, all the standard (U)HPLC components, for example, autosampler, column compartment/oven, and detectors are deployed with modifications necessary for handling a fluid mixture of CO_2 and a cosolvent. The entire system is maintained at an elevated pressure by an automated backpressure regulator (ABPR) to keep the mixture as a single, homogeneous phase. This demonstrates that

Abhijit Tarafder, Steven M. Collier, Jason F. Hill, Waters Corporation

https://doi.org/10.1515/9783110500776-003

the challenges related to designing a robust SFC system that can constantly generate repeatable results originate from the use of CO_2 – especially the high compressibility feature of CO_2. Even in liquid form, CO_2 is about three times more compressible than methanol and more than five times compressible than water. Apart from this physical aspect, presence of CO_2 along with near absence of water in the mobile phase of SFC system facilitates reactions that are not normally encountered in LC.

The chapter is divided into three major sections. First, a brief overview on the effects of CO_2 compressibility on various system components and how it challenged instrument design will be provided. We will then discuss how run-to-run reproducibility with the same column and the same system can be compromised over the time in SFC. Also, how CO_2 compressibility amplifies batch-to-batch variations in column performance and the measures that were taken to address these issues. Lastly, we will demonstrate how even after designing the most robust instrument modules and columns, SFC as a system may fail in robustness and repeatability tests, unless special measures are taken.

3.2 Effect of compressibility on instrument robustness

3.2.1 Pump

Pumping CO_2 presents several challenges. Standard CO_2 supply comes in CO_2 tanks, where liquid CO_2 at ambient temperature is available. The space above liquid CO_2 is filled with CO_2 vapor whose pressure depends on the ambient temperature. Reciprocating pumps, which are normally employed in commercial instruments, are not designed to work with CO_2 vapor. Standard practice is to either draw liquid CO_2 from the tank using a dip-tube or cool the incoming CO_2 to a set condition, or both. The ratio between the constant pressure (C_P) and constant volume (C_V) heat capacities of CO_2 is much higher than standard liquid solvents, which means the process of compressing CO_2 while pumping can lead to significant rise in CO_2 temperature unless the pump-head temperature is maintained at the set point by a very efficient cooling mechanism. As liquid CO_2 is about three to five times more compressible than standard (U)HPLC solvents, the mass of CO_2 delivered by the same volumetric flow from a pump will depend on the set pump-head temperature and also on the efficiency of pump-head cooler.

Single-stage repurposed reciprocating LC pumps are inadequate in compressing and delivering CO_2 repeatedly and reliably. Because of the reasons mentioned earlier, these pumps caused variability in mobile-phase mass flow rates as well as mass compositions. Such variations impart changes in the solvation power of mobile phase and often results in shifting retention times from injection to injection or system to system. Modern SFC systems deploy two-stage reciprocating pumps. Plunger movements of each stage are precisely controlled for the intake, compression, and delivery strokes. Often the pump heads are independently and very efficiently cooled, improving density control of CO_2 and hence accurate mass delivery.

3.2.2 Sample injection

In many earlier SFC systems only full loop injections were possible because with partial loop injections it was difficult to maintain injection solvent homogeneity, leading to lesser accuracy, precision, and injection linearity. In addition, large volumes of sample were wasted with every injection, or, the users were forced to change the sample loop manually whenever necessary. These restrictions limited system flexibility. Modern systems are equipped with dual-injection valves. The valve design allows the primary sample loop to vent to the waste, enabling the sample to enter the loop under atmospheric pressure while maintaining the homogeneity of the mobile phase. In addition, the auxiliary injection valve reduces pressure pulses from the injection event and also addresses the carryover issue, which enables reproducible and reliable partial loop injections.

3.2.3 Optical detection

Compressibility of CO_2 can severely affect the refractive index (RI) of SFC mobile phase, which in turn can create significant baseline noise and curvature. Commonly used solvents in (U)HPLC, for example, acetonitrile (RI = 1.34), methanol (RI = 1.33), and water (RI = 1.33) [1] have very similar RI values, and so the effect of RI variation in (U)HPLC methods is negligible. Carbon dioxide, on the other hand, has a very different RI value compared to methanol – the most common cosolvent in SFC. In addition, the variance of RI value of CO_2 as a function of density is nonnegligible. For example RI of CO_2 is 1.14 at 0.62 g/mL and 1.16 at 0.68 g/mL [2]. Although not too high, this difference in RI values is sufficient enough to increase baseline noise if the density variation in the flow cells is not closely controlled. As baseline noise directly affects sensitivity, this issue is very important.

Modern SFC detectors are specially designed for compressible fluids and working at high pressures. For example, instead of sapphire lenses that reduce energy throughput at lower ultraviolet wavelengths, modern detector's lenses are made of high-strength silica that can withstand the high pressure of the system. Methods are employed to minimize density variation inside the flow cell. Optics bench has thermal control to mitigate density fluctuations. These measures improve baseline stability and mitigate RI effects. Low-dispersion flow cells are employed to minimize extra-column dispersion.

3.2.4 Backpressure regulation

Backpressure regulation (BPR) in SFC can be difficult because of the high compressibility of CO_2. The mobile phase expands dramatically across the valve over a length of few microns which leads the expanding fluid to flow close to the speed of sound, creating acoustic shocks. In addition, followed by a Joule–Thomson cooling, expanding CO_2 directly transforms into solid CO_2 particles through deposition. The impacts generated on the pressure regulation point by these phenomena often lead to rapid degradation of constituent materials. Along with these physical challenges, earlier SFC systems suffered from poor pressure monitoring, slow response time to feedback loops, and low-resolution stepper motors.

Modern SFC systems exhibit improved ABPR control by employing high durability material and dual-stage pressure control. Normally a sacrificial, static backpressure control is employed to spare the main pressure regulator from severe pressure change conditions. The static BPR maintains the system at a minimum allowable pressure while the active BPR controls the user-defined set point. To address the deposition issue, the static cartridge BPR is often heated to a high temperature. All the steps in the pressure control system are made significantly more efficient for faster and precise response to disturbances.

3.3 Addressing the robustness and reproducibility of achiral SFC columns

Maintaining chemical and mechanical stability of the chromatographic column is equally important for providing reproducible results with repeated injections throughout the life of the column. There are a multitude of factors that influence robustness and repeatability of column performance. This section will focus on two areas that are unique to SFC column design.

1. Modification of surface chemistry – during SFC experiments retention and selectivity of traditional silica-based SFC columns can change slowly over time while in use, raising robustness concern [3, 4]. Modern SFC columns have been designed to mitigate or eliminate such retention drift.
2. Contribution of column bed density – while using compressible CO_2/cosolvent mobile phases, there is a need for SFC columns to be packed with consistent column-to-column "backpressure" or consistent, predictable pressure drop across the length of the column, to insure reproducible separations. Special measures are now taken to pack and test SFC columns to reduce variability.

In the following sections these two factors will be discussed in detail.

3.3.1 Modification of surface chemistry of silica-based SFC columns

Typical achiral SFC columns are based on silica and the chromatographic silica particle surface is composed of silanols. Silanols are strongly hydrophilic in nature and interact with polar functional groups of mobile phase and analyte compounds. SFC generally operates in normal phase mode where the retention mechanism is mostly controlled by polar–polar interactions between analytes and the stationary phase chemistry. To provide different selectivity and better manage peak shape, SFC columns are manufactured with different surface modifications or bondings to provide different levels of polarity. Popular SFC column bondings include 2-ethylpyridine, 4-ethylpyridine, *para*-fluorophenyl (PFP), cyano, aminopropyl, and diol. Changes to the polarity or chemical nature of column surfaces affect retention and selectivity. For example, any phenomenon that leads to decreasing the polarity of particle surface should result in loss of retention and possibly selectivity changes. Formation of silyl ether from the reaction between silanol and methanol is such a phenomenon, which will be discussed in this section.

3.3.1.1 Effect of silyl ether formation

The issue of gradual loss of retention over SFC stationary phases has been known for some time. For example, a methodical study from Bristol-Myers Squibb clearly demonstrated [4] that with silica-based columns, when used for SFC applications, selectivity and retention times can change with time (see Figure 3.1). Detecting the real reason behind such retention changes, or retention drift, can be challenging because a combination of several other factors may lead to similar variations in retention. For example, a leak from repurposed CO_2 pump can increase over time, causing increasing composition of modifier in the mobile phase, leading to decreasing retention. Other factors like structural failure of packed beds, for example, generation of voids or bed shifting, can lead to similar consequences.

Methodical research backed by robust instrumentation helped to focus on the possibility of modification of the surface chemistry of silica-based columns that can occur under SFC conditions [3]. It was detected that in water-free compressed CO_2/methanol mobile-phase environment, methanol reacts in a condensation reaction with the surface silanols to form silyl ethers (see Figure 3.2). This, in turn, changes the polarity of the silica surface, which impacts changes in retention and selectivity.

Silyl ether condensation reaction requires three basic factors: (1) the liquid phase must be nearly water free, (2) the liquid phase must contain alcohol, and (3) for silanol and alcohol to condense, the subunits must be in proximity. All three factors are present in an SFC system. As the slow conversion from silanol to silyl ether occurs, polarity of the silica-based particles reduce and chromatographers observe drift to shorter retention times and selectivity changes over a period of days to weeks (see Figure 3.3(a)). Mobile-phase additives can further increase the retention drift and selectivity changes

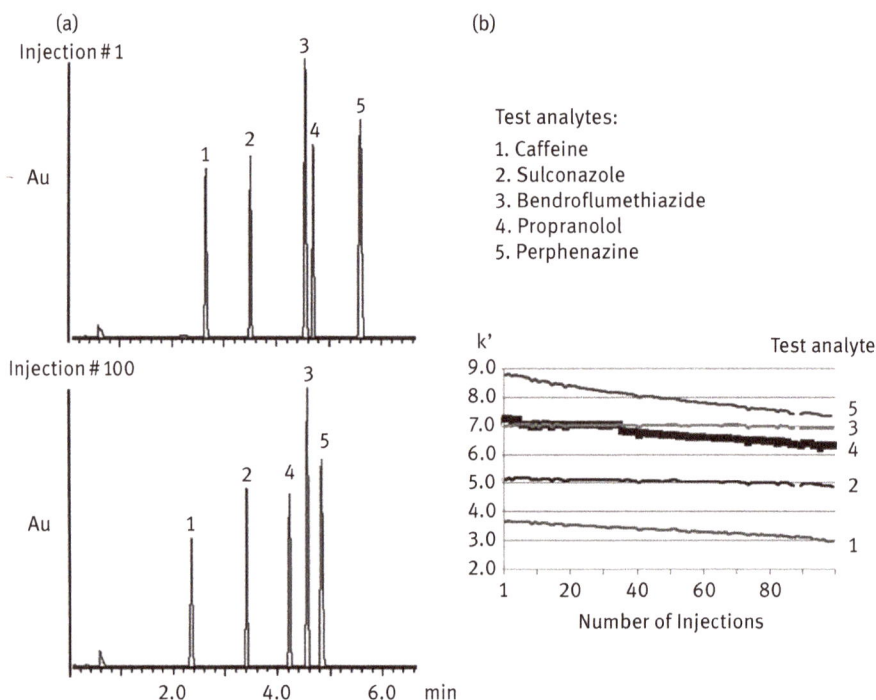

Figure 3.1: Example of gradual retention loss of basic analytes over 100 injections. Note that at the end the elution order of propranolol and bendroflumethiazide had reversed. Reprinted with permission from Ref. [4].

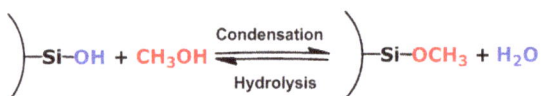

Figure 3.2: Formation of methyl silyl ether from silanol and methanol on a silica-based particle. Reprinted with permission from Ref. [1].

in some cases. For example, mixing a tertiary amine into methanol cosolvent accelerates silyl ether formation on silica-based columns with water-free CO_2/methanol mobile phase [3]. Both chromatographic and nuclear magnetic resonance studies confirmed the theory of formation of silyl ethers on both bonded (2-ethylpyridine) and unbonded silica (and inorganic/organic hybrid silica) particle surfaces (see [3]).

3.3.1.2 Mitigating silyl ether formation

Identification of silyl ether formation and the ensuing effect on reproducibility demanded ways to mitigate the effect. Other particle substrates were investigated.

For example, bridged-ethyl hybrid (BEH) particles [5], which have lower amounts of silanol groups as compared to silica particles, reduce the amount of retention drift by about 50%. Although hybrid particles have only ~10% lower silanols compared to silica, the near two-times reduction in the retention drift is mainly caused by lower silanol activity, not due to the lower number of silanols on BEH surface. Another avenue is to regenerate SFC column through hydrolysis to regain the original retention factors. Experimentally it was demonstrated that columns exhibiting lower retention due to silyl ether formation could be returned near to its initial retention state by flushing with water [3] (see Figure 3.3(b)).

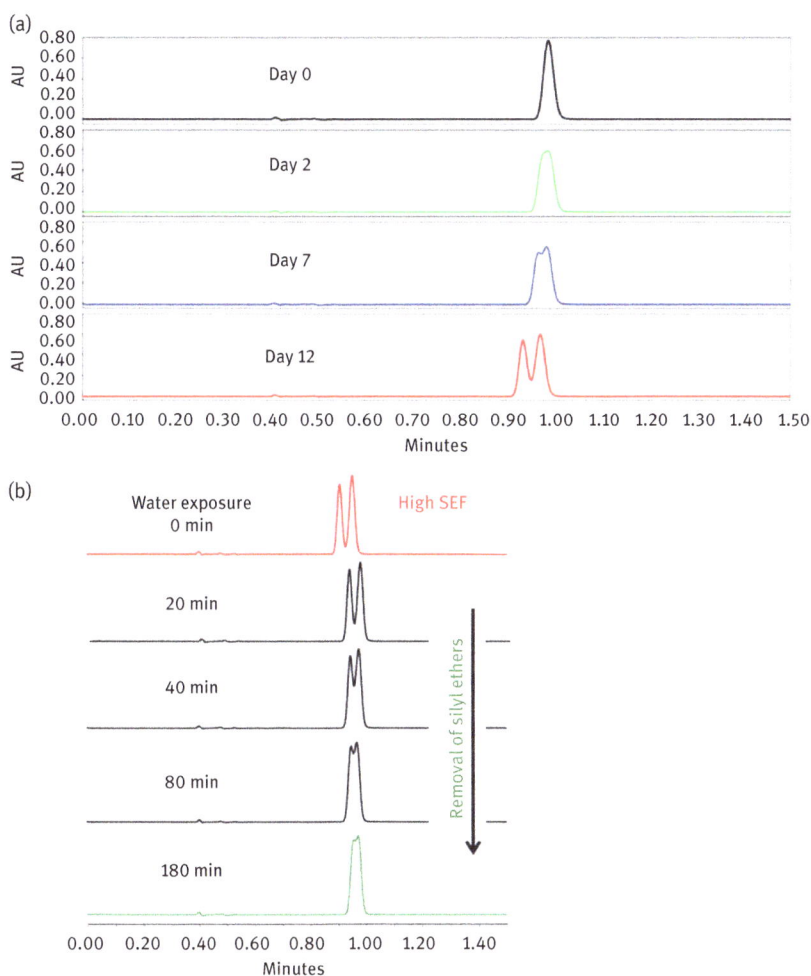

Figure 3.3: (a) Demonstration of surface modification due to silyl ether formation. 3-Benzoylpyridine and flavone change selectivity when the column was stored in pure methanol between data points. (b) Rinsing the same column with water removes the silyl ether and can regenerate the column. Chromatographic conditions for both (a) and (b) are the same. Reprinted with permission from Ref. [1].

Although these are workable fixes, the situation demanded a permanent solution. To address silanol reaction at the surface, synthetic chemists of Waters Research and Development looked for ways to modify the surface for either mitigating or eliminating silyl ether formation. A series of prototype columns were packed and evaluated. These prototypes needed to address robustness by eliminating the silyl ether retention issue, while also providing high levels of SFC chromatographic performance. As a result, Waters' Torus family of columns was developed that is based on a patent-pending two-stage functionalization of BEH particles [6].

In Torus columns, the initial bonding provides a hydrophilic surface that controls the retention characteristics of the sorbent and is responsible for minimizing unwanted surface interactions that lead to retention and selectivity changes over time (see Figure 3.4). The second step of the functionalization is responsible for individual selectivity and peak shape characteristics of each of the Torus chemistries (see Figure 3.5).

Following the functionalization step, four SFC column chemistries were developed to provide added selectivity over a wide range of compounds:

– Torus 2-PIC (or 2-picolylamine) is primarily for basic compounds with and without the use of additives (see example).
– Torus DEA (or diethylamine) has different selectivity compared to 2-PIC and exhibits good peak shape with basic analytes.
– Torus DIOL is a high density diol for acidic analytes.
– Torus 1-aminoanthracene is for hydrophobic or neutral compounds such as vitamins, lipids, and steroids.

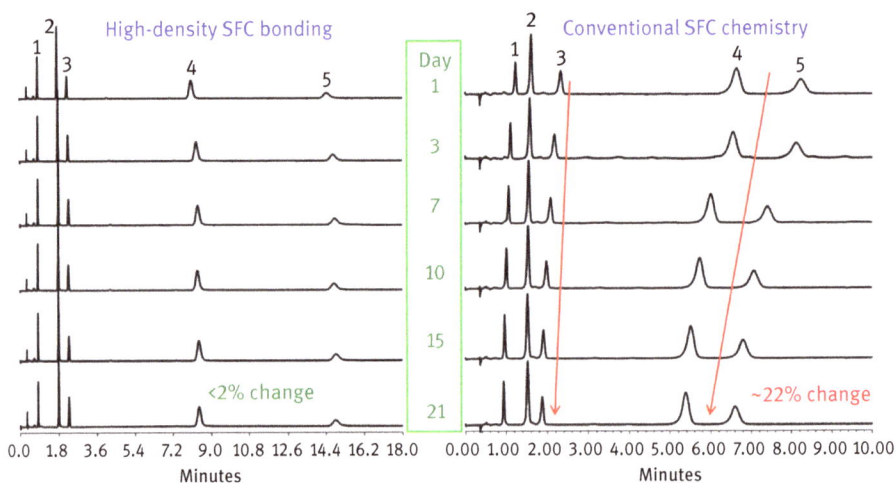

Figure 3.4: Example of mitigation of silyl ether formation followed by a two-stage functionalization of ethylene bridged hybrid (BEH) particles. Reprinted with permission from Ref. [6].

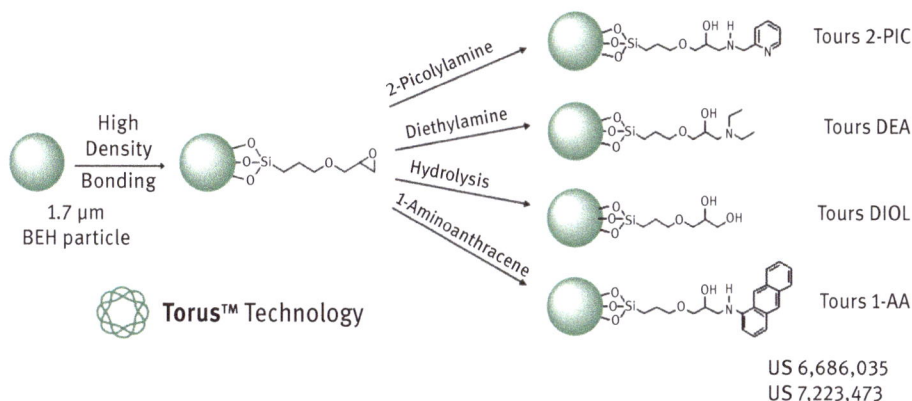

High
Density
Bonding

1.7 µm
BEH particle

2-Picolylamine

Diethylamine

Hydrolysis

1-Aminoanthracene

Torus™ Technology

Tours 2-PIC

Tours DEA

Tours DIOL

Tours 1-AA

US 6,686,035
US 7,223,473

Figure 3.5: Column chemistries implemented to mitigate silyl ether formation and ensuing retention drift with Torus family of SFC columns.

Studies show that Torus columns can be used with common acidic and basic additives such as trifluoroacetic acid, formic acid, ammonium acetate, ammonium formate, ammonium hydroxide, and organic amines (such as DEA and triethylamine) and ammoniated methanol, up to a concentration of 20 mM or 0.2%. Suggested additives against each of the chemistries are listed in Table 3.1.

Table 3.1: Suggested additives for different Torus chemistries.

Suggested additives for Torus column chemistries	20 mM NH_4OH (ammonium hydroxide)	20 mM NH_4Ac (ammonium acetate)	0.2% TFA (trifluoroacetic acid)
Torus 2-PIC (2-picolylamine)	✓	–	✓
Torus DIOL (high-density diol)	–	✓	✓
Torus DEA (diethylamine)	✓	–	–
Torus 1-AA (1-aminoanthracene)	✓	✓	✓

3.3.2 Effect of packing density on column-to-column reproducibility

Instances of surface modification during chromatographic run are not unique to SFC – it is encountered in (U)HPLC as well. For example, in normal-phase liquid chromatography there are equilibration issues, often due to the presence of residual water. Also strong retention of certain additives on (U)HPLC columns leads to the so-called memory-effect. The effect of column packing on retention reproducibility, however, is unique to SFC. In SFC, between two columns with the same surface properties, any change in the pressure drop or the backpressure along the column can impact the separation under certain method conditions. When experiments are conducted under pressure–temperature conditions where the mobile phase is more compressible, variations in backpressure lead to stronger variations in density along the column, resulting in changes in chromatography.

Such situation makes the challenges in column packing more difficult compared to the situations encountered in (U)HPLC. Typically, as (U)HPLC columns are packed, the resultant backpressure of the finished columns are monitored only for extreme changes because any changes in backpressure have nominal impact due to the essentially incompressible nature of the liquid mobile phase. For packing SFC columns, however, a more stringent criterion needs to be maintained to avoid column-to-column irreproducibility. The example shown in Figure 3.6 demonstrates the effect of different backpressure on the chromatography. Different chromatograms resulted as a result of different column backpressures packed with the same batch of media, and while keeping the method conditions and the SFC system the same.

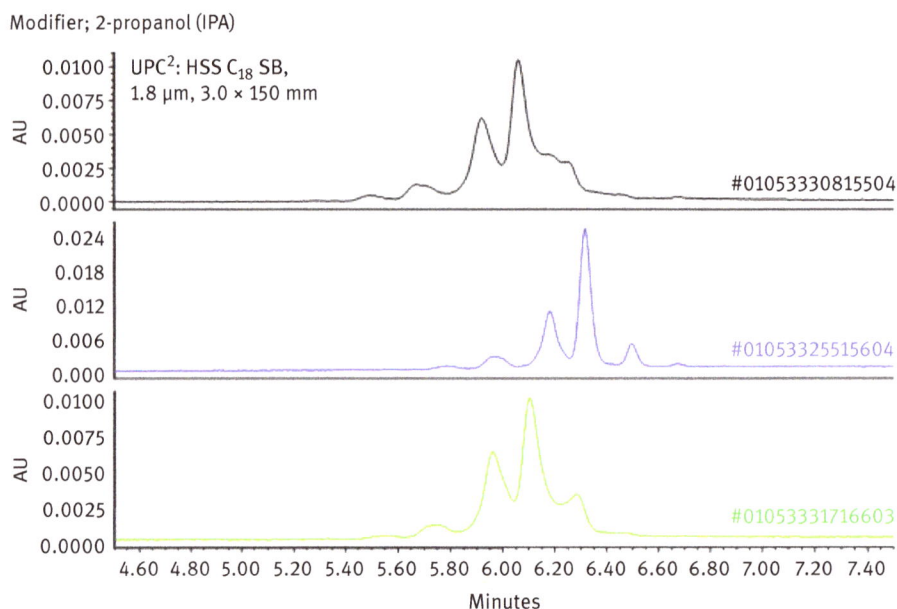

Figure 3.6: Column-to-column variability between different columns packed with the same batch of chromatographic media. These results exemplify the critical role played by column packing in influencing chromatographic retention in SFC.

To ensure reproducibility, related slurry-packing protocols were developed and replicated at all the facilities, which led to uniform and scalable column backpressure.

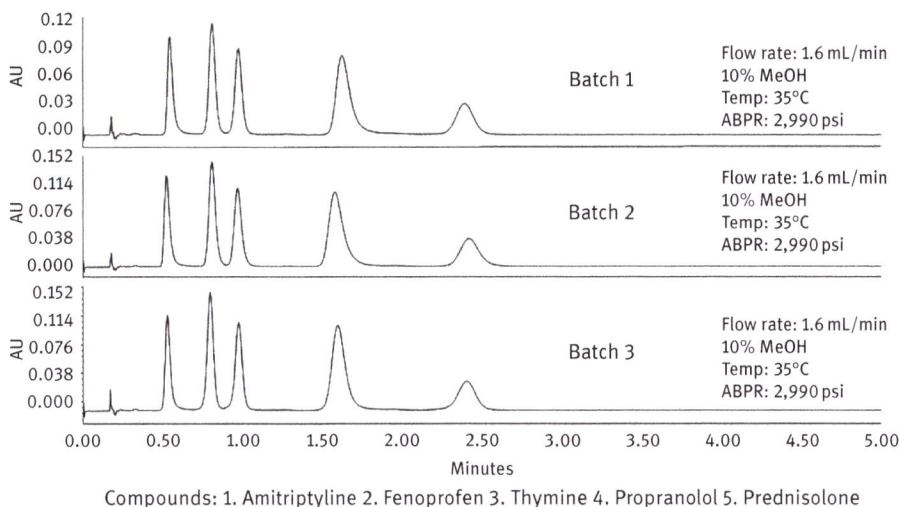

Compounds: 1. Amitriptyline 2. Fenoprofen 3. Thymine 4. Propranolol 5. Prednisolone

Figure 3.7: Column-to-column repeatability ensured after implementation of stringent packing protocol with SFC columns. Experiments carried out with a 3 × 50 mm Torus 2-PIC column packed with 5 μm particles.

Figure 3.7 demonstrates how a stricter column packing regime leads to batch-to-batch reproducibility. Similar protocol also led to better column scalability. For example, moving from 50 mm length to 75 to 100 mm SFC columns, the transfer may confront scalability issues due to inconsistent variation of backpressure. A common concern from the practitioners is that each column length should be packed with a resultant backpressure that is "predictable" or scalable as one move to shorter or longer length columns. Protocols were developed to maintain a narrow pressure drop window per length of column, as measured in psi/mm, so that the backpressure of a 100 mm column is approximately twice the backpressure of a 50 mm column.

   With modern SFC columns, such as the Torus column family, all column chemistries of a particular length, ID and particle size are produced within a narrow backpressure specification range. This offers two advantages which includes (1) column-to-column reproducibility, and (2) the ability to establish SFC column screening methods for a selection of different column chemistries without the need to adjust method conditions when switching between columns. It further enables the ease of scale-up to purification.

3.4 Effect of system variability on instrument robustness

SFC systems provide ideal example of demonstrating how a combination of error-free components may not always lead to a perfect system. The reason again is the compressibility of SFC mobile phase. The ABPR which is deployed to ensure mobile-phase homogeneity by maintaining the system at high pressures – generally more than 100 bar – is placed at the very end of the system. The controller of ABPR valve works on a feedback loop, sensing the pressure immediately upstream to the valve opening, and trying to maintain that pressure only at the set point supplied by the user. Although the set pressure can be precisely maintained at the upstream of ABPR, there is no direct control on the pressure, and hence the density, further upstream of the system, which means any change in method conditions or column selection that can lead to changing resistance along flow path of the mobile phase will lead to changes in pressure along the system, including pressure inside the column. While this is true even for (U)HPLC systems, the consequence of pressure variation in SFC can be significant compared to the LC counterparts. Due to higher compressibility of SFC mobile phases, having different pressure profiles along the column caused by different flow resistances can lead to different density variations and hence variations in chromatography. Understandably, such uncontrolled variation in pressure, upstream of the ABPR will have consequences on system robustness and repeatability.

Note that the consequences will depend on the compressibility of the mobile phase and also the nature of the analyte. With high concentration of liquid modifiers in the mobile phase or at low temperatures and high pressures, SFC mobile-phase compressibility is low. So, under such conditions we may not have noticeable effect on the retention variations due to pressure changes. Under high compressibility regions, on the other hand, robust operations may be compromised. A detailed discussion on the effect of experimental condition selection on compressibility is provided in Section 4.1. Also note that even for method conditions where mobile-phase compressibility is high and its effect on system robustness is nonnegligible, there are ways to mitigate the problem. A discussion on this solution is provided in Section 4.2.

3.4.1 Effect of method condition on system robustness

A case study published earlier [7] demonstrates how choice of method condition can affect method repeatability. The study conducted a robustness analysis which is carried out to study the extent of chromatographic variation that can be caused by some unintentional variation of method parameters during execution. Such exercises are often conducted while validating methods across different instruments and between different laboratories.

Within the operating space of commercially available SFC instruments, that is, within the high and low limits of pressures and temperatures allowed, and also depending on the composition, SFC mobile-phase compressibility can significantly vary (see Table 3.2).

As a rule of thumb, SFC mobile-phase compressibility decreases as we increase pressure or decrease temperature. Compressibility also decreases when the cosolvent percent is increased and vice-versa. For example, if we have a method set at low

Table 3.2: Variation of solvent compressibility of CO_2 and CO_2 + methanol mixtures at different pressures and temperatures.

Mass frac. (carbon dioxide)	Mass frac. (methanol)	Temperature (°C)	Pressure (bar)	Density (g/cm³)	Adi. compress. (1/bar)
1	0	30	175	0.87079	0.000401
1	0	60	175	0.67638	0.001108
1	0	30	300	0.94798	0.000234
1	0	60	300	0.82971	0.000405
0.5	0.5	30	175	0.84924	0.000104
0.5	0.5	60	175	0.80418	0.000147

pressure, high temperature, and low cosolvent composition, we expect to experience stronger variation in mobile-phase density even with nominal changes in method parameters because of high mobile-phase compressibility. But at higher pressure and lower temperature, especially with high percent of cosolvent, density variation is minimal because of decrease in compressibility. At a fixed mobile-phase composition, SFC retention is mainly driven by the density of the mobile phase, so the method robustness is controlled by mobile-phase compressibility.

Three example situations were considered to carry out robustness analysis. The control mobile-phase composition was chosen as 5% methanol. Although SFC solvent gradients reach much higher percent of cosolvent, many times the gradient starts at 5% or below, which was why 5% was chosen for this study. Details of the method conditions are provided below:

Method 1
-   Flow rate: 1.2 mL/min ±4%
-   Cosolvent %: 5% methanol ±4%
-   Column temperature: 55 °C ±10%
-   Backpressure: 1,750 psi ±10%

Method 2
-   Flow rate: 1.2 mL/min ±4%
-   Cosolvent %: 5% methanol ±4%
-   Column temperature: 55 °C ±10%
-   Backpressure: 3,500 psi ±10%

Method 3
-   Flow rate: 1.2 mL/min ±4%
-   Cosolvent %: 5% Methanol ±4%

–   Column temperature: 25 °C ±10%
–   Backpressure: 3,500 psi ±10%

Note that for each method, the flow rate, the cosolvent percent, the column temperature, and the ABPR pressure were varied from the control condition. Figure 3.8 provides a graphical view of the locations of the method parameters on a pressure versus temperature plane. The blue contour curves shown in the figure are the constant density, or isopycnic curves of 95/5(mol/mol, %) CO_2 + methanol mixture. The curves are plotted with densities at equal numerical intervals which can be used to understand fluid compressibility as well based on the spatial proximity of the curves. For example, within the method parameter variation box of Method 1, the number of curves passing is the maximum (see Figure 3.8) followed by the boxes of Methods 2 and 3. This indicates that mobile-phase compressibility around Method 1 is the most, followed by Methods 2 and 3. And hence, consequences of method parameter variation should be the most for Method 1 followed by the other two in that order.

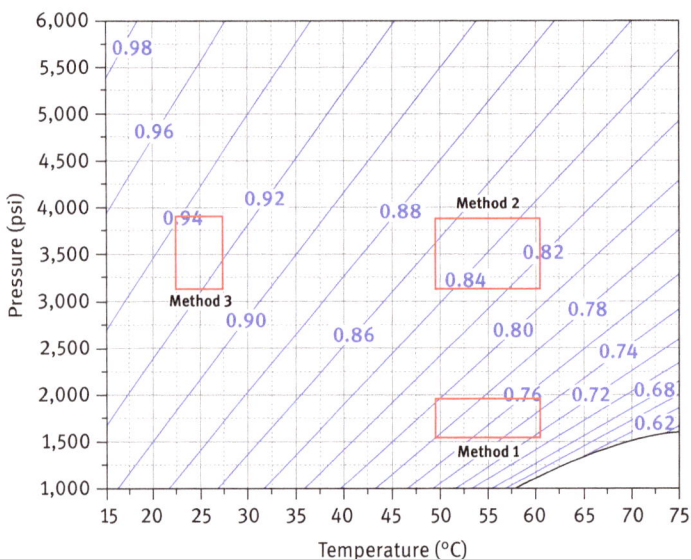

Figure 3.8: Operating regions of Methods 1–3 described in Section 4.1. Blue curves on the pressure–temperature plane represent constant density or isopycnic plots of 95/5 (mol/mol, %) CO_2 + methanol mixture. Densities in g/mL corresponding to each curve are presented on the figure as blue numerical values. The operating space of parameter variation around each method is represented as red boxes, where the pressure and the temperature were varied by 10% for each of the three methods. Reprinted with permission from Ref. [7].

Figure 3.9 demonstrates the chromatographic variations with 4% changes in the flow rate and cosolvent percent, and with 10% changes in the temperature and the ABPR pressure. The variations imparted to this method resulted in an average of 10.1%

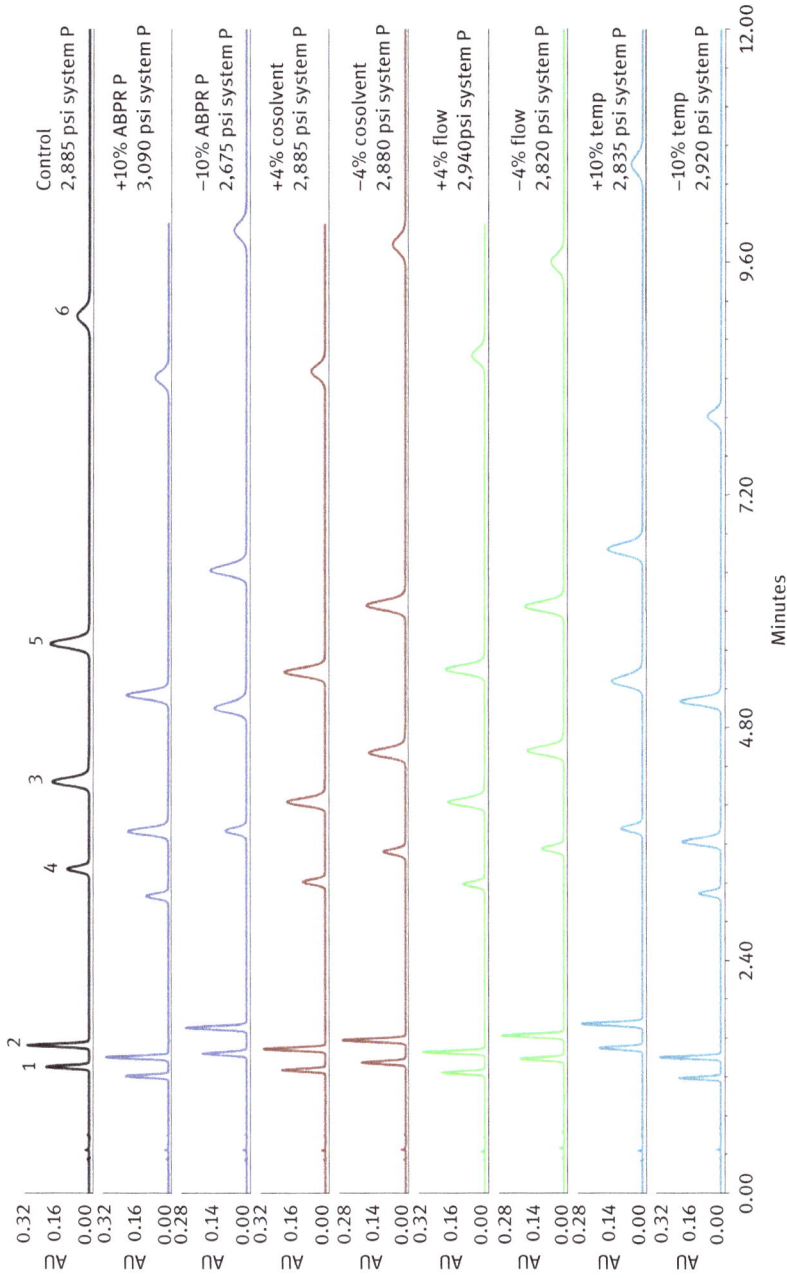

Figure 3.9: Chromatograms demonstrating the effect of parameter variation around Method 1 described in Section 3.4.1. The control method conditions are 1.2 mL/min flow rate, 5% methanol, 1,750 psi ABPR, 2885 psi system pressure and 55 °C. Analytes are caffeine (1), flavone (2), papaverine (3), guaifenesin (4), carbamazepine (5), and umbelliferone (6). Reprinted with permission from Ref. [7].

difference in retention factors for the analytes. For Method 2, which has the control ABPR pressure set at 3,500 psi compared to 1,750 psi in Method 1, the average difference in retention factor is reduced to 5.5% (see Figure 3.10). This improvement in robustness could be achieved because Method 2 is located in a lesser compressible zone.

Figure 3.11 demonstrates the chromatograms corresponding to the robustness analysis around Method 3. Here even lesser variation in chromatography can be observed (average difference in retention factor is 2.6%) since Method 3 is located in an area of very low compressibility. Figure 3.12 compares the changes in retention factors for all the three method conditions discussed earlier. In general, the trend shows progressively lesser chromatographic variation, in response to the same variations in method parameters, at regions that have lesser mobile-phase compressibility.

Note that, although all the factors varied in the examples provided here imparted variations in chromatography, the physical nature of impact by the flow rate variation is different from the nature of impact of variations in cosolvent, pressure, and temperature. Flow rate has an indirect impact on the mobile-phase density. In (U) HPLC, changing flow rate do not affect retention factor, but in SFC it does. In SFC the changes in system pressure caused by the change in flow rate changes mobile-phase density and therefore changes analyte retention factors. Also note that the percent cosolvent variation generally has maximum impact on analyte retention compared to other factors.

In summary, the results shown here demonstrate (a) how mobile-phase compressibility can vary within the standard operating space of an SFC system, and (b) how mobile-phase compressibility affect system robustness even when with robust system components.

3.4.2 Methods to mitigate system robustness issues

As discussed earlier, SFC system robustness issue originates from uncontrolled variation of upstream pressure/density conditions which can be caused by variations in method conditions or while repeating the method in another column or in another system. A convenient solution to this problem, which is not perfect but should be sufficient enough under most practical situations, is to shift the point of pressure control from the very end of the system to the middle of the system – preferably at the middle of the column. Figure 3.13 demonstrates this concept schematically. This shift in pressure control can be implemented either manually or through automated means where the proportional-integral-derivative (PID) controller of the ABPR controls the immediate ABPR upstream pressure in such a way that the average of the measured pump outlet and the ABPR pressures is controlled, in place of the ABPR pressure itself.

Figure 3.13 illustrates the situations described earlier. In Figure 3.13(a), variations of pressure along an ABPR-controlled system under three different conditions are

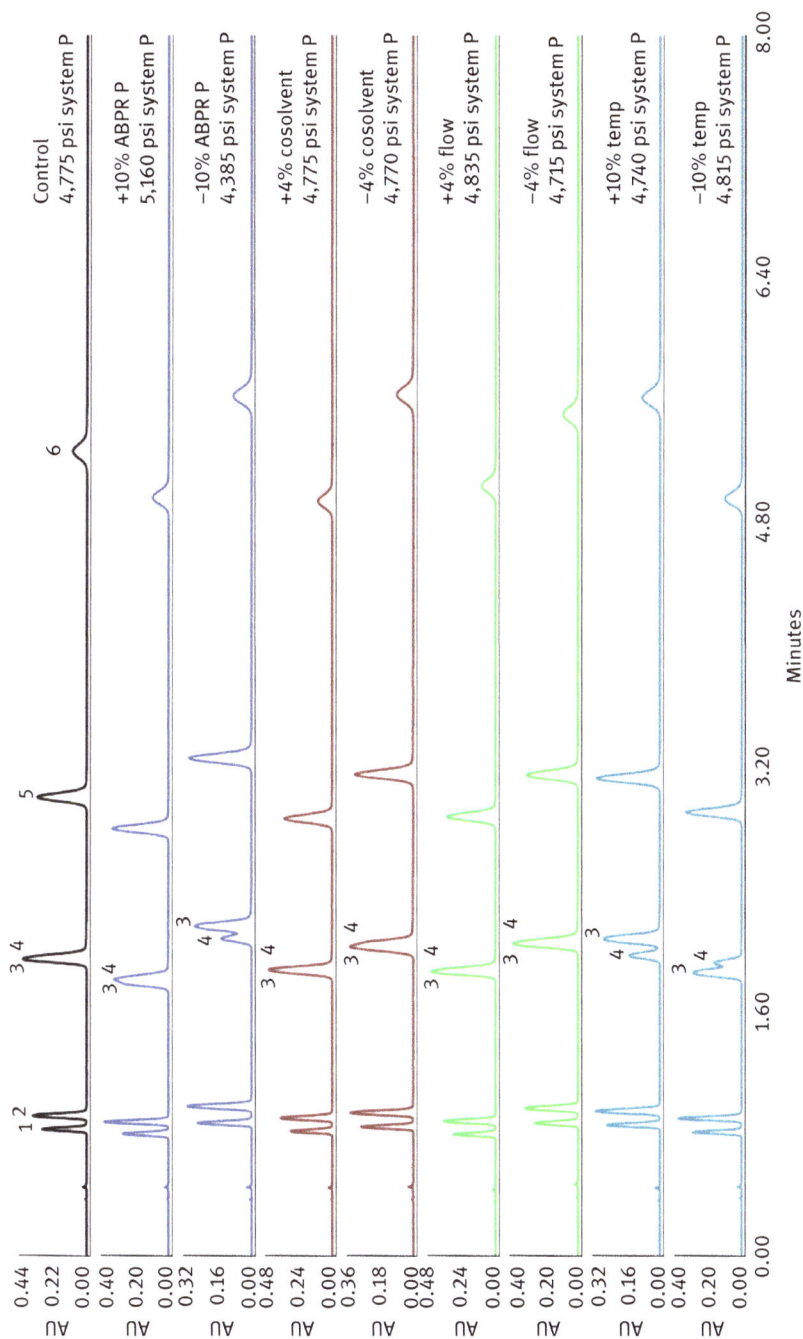

Figure 3.10: Chromatograms demonstrating the effect of parameter variation around Method 2 described in Section 3.4.1. The control method conditions are 1.2 mL/min flow rate, 5% methanol, 3,500 psi ABPR, 4775 psi system pressure and 55 °C. Analytes are caffeine (1), flavone (2), papaverine (3), guaifenesin (4), carbamazepine (5), and umbelliferone (6). Reprinted with permission from Ref. [7].

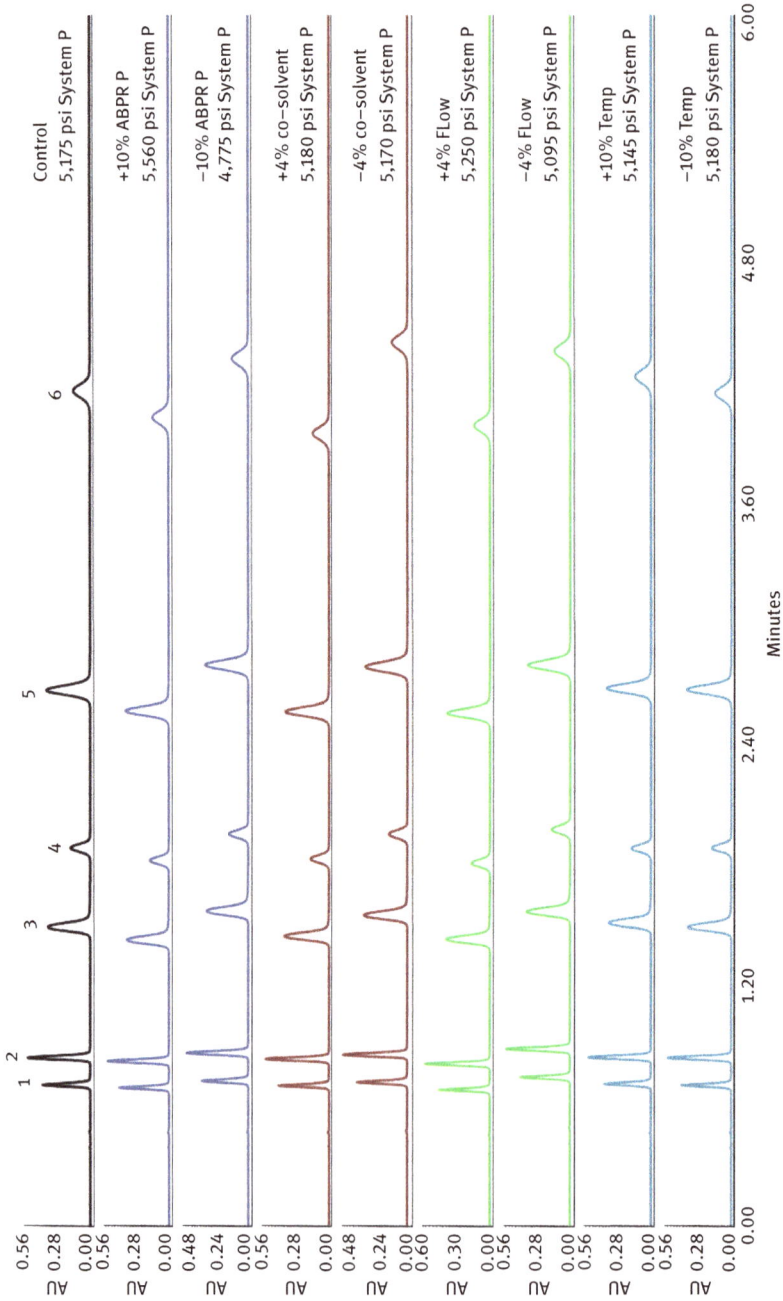

Figure 3.11: Chromatograms demonstrating the effect of parameter variation around Method 3 described in Section 3.4.1. The control method conditions are 1.2 mL/min flow rate, 5% methanol, 3,500 psi ABPR, 5175 psi system pressure and 25 °C. Analytes are caffeine (1), flavone (2), papaverine (3), guaifenesin (4), carbamazepine (5), and umbelliferone (6). Reprinted with permission from Ref. [7].

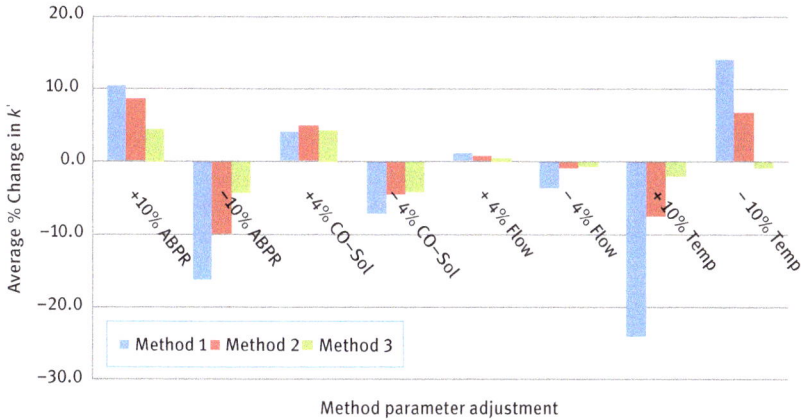

Figure 3.12: Average percent variation in retention factors imparted by changes in different parameters around Methods 1, 2, and 3. Reprinted with permission from Ref. [7].

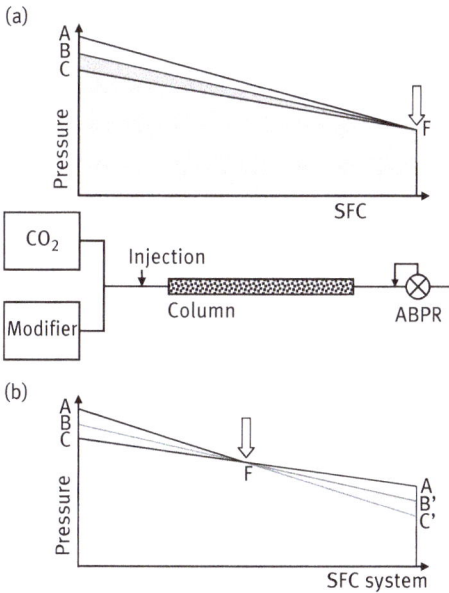

Figure 3.13: (a) Variation of pressure profile along an SFC system when the pressure is regulated at the end. Any variation in flow resistance either increases or decreases the pressure profile over the entire system. F here denotes the ABPR pressure which is always maintained constant in this mode of operation. B is the pump outlet pressure of a reference condition, and A and C are the pressures at the same point with higher and lower flow resistances, respectively, over the system. (b) Variation of pressure profile when the pressure is maintained at the middle of the system. Any variation in flow resistance will still make the pressure profile either steeper or gentler, but the changes at the upstream of the middle point F are mitigated by the changes at the downstream. F here denotes the average pressure of the system. Pressures B and B' represent the pump outlet and the ABPR pressures, respectively, of a reference condition. Profiles AC' and CA' represent conditions when the flow resistances were higher and lower, respectively.

demonstrated. A linear pressure variation is considered for the sake of simplicity – actual pressure profiles along the system can be different. Line BF in Figure 3.13(a) represents pressure profile at the reference method condition. If there are any variations in flow resistance – caused either by variations in method condition or column permeability or system plumbing – the pressure profile can move either toward higher or lower pressures, depending on higher or lower flow resistances, respectively. Such situations are represented by lines AF and CF, respectively, in Figure 3.13(a). If such pressure variations lead to sufficient variation in density and if the analytes are sensitive to such density variation, changes in retention and selectivity will result because throughout the elution path analytes will experience either increased or decreased retention.

On the other hand, if the pressure/density at the middle of the system is controlled, a change in flow resistance will create different pressure profiles, like the example in Figure 3.13(a), but the conditions at the upstream of the middle point (see point F in Figure 3.13(b)) can be mitigated by the conditions downstream of it. The situation is explained in Figure 3.13(b). Note that, in this figure, the reference situation is represented by the pressure line BB′. When there is an increased resistance, the ABPR pressure is reduced to C′ either manually or automatically, to maintain the same set average pressure of F. The new pressure profile is represented by the line AC′. Similarly when there is a decreased resistance, ABPR pressure is increased by the controller to A′ and the pressure profile is represented by CA′. If we assume that retention factors of analytes have a linear dependence on mobile-phase density, and the density has a linear dependence on pressure, an increased or decreased retention at the first half of the journey of an analyte through the column will be compensated during its journey through the second half. Detailed explanation on this issue can be obtained at Refs. [8, 9].

An example of the applicability of this approach is presented in Figure 3.14. Figure 3.14(a) demonstrates a chromatogram obtained with a 2.1 × 150 mm column filled with 1.7 μm particles. Six compounds present in the sample mixture were well separated in this example. When a column with same dimensions but filled with 5 μm particle is deployed, while keeping all the other experimental conditions and the SFC instrument the same, we note a drastic change in chromatography (see Figure 3.14(b)). The peaks are broader here, which can be expected by an (U)HPLC practitioner because of the increase in particle size, but the selectivity of separation also changed in this example – which is unexpected. Figure 3.14(a) demonstrates how even with the same system and the same method conditions, the approach of controlling pressure at the end of the system leads to very different density profiles across the system when the columns are packed with different particle sizes. The situation can be considerably mitigated when the control point of the pressure is shifted to the middle of the system (see Figure 3.14(b)). In this case, although the density profiles with the columns with two different particle sizes are very different, their average conditions match (see Figure 3.14(b)). Such matching, although not perfect, notably improves system robustness which can be observed from the chromatograms presented in Figure 3.14(c).

(1) Caffeine, (2) carbamazepine, (3) uracil, (4) hydrocortisone, (5) prednisolone, and (6) sulfanilamide

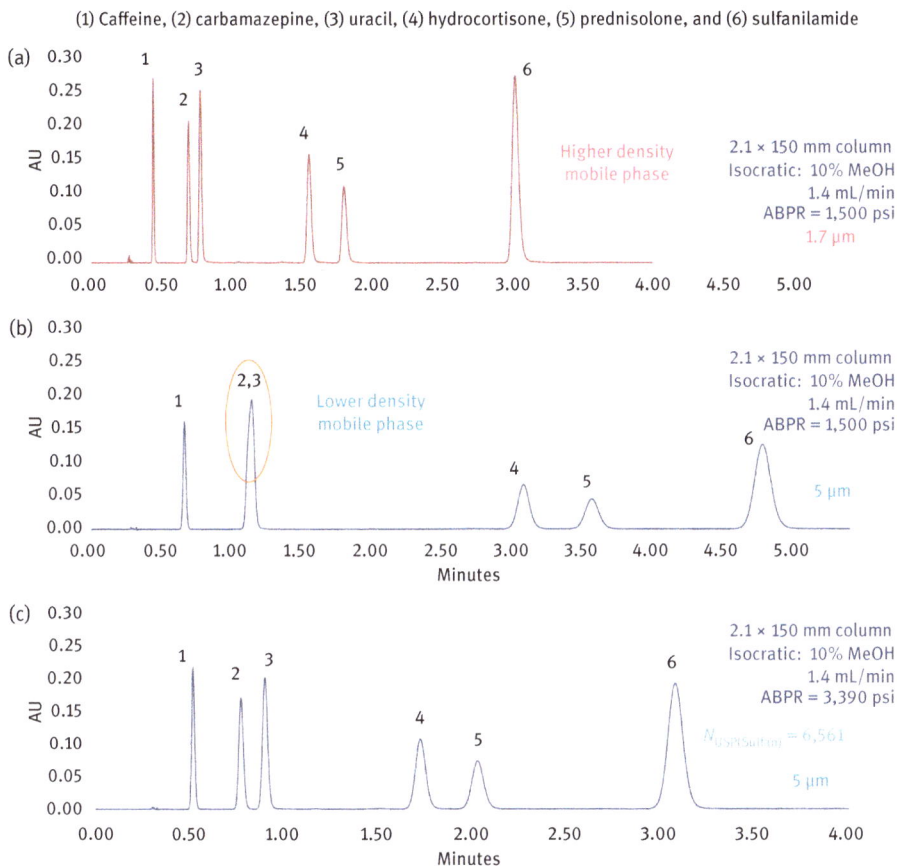

Figure 3.14: (a) Chromatogram obtained with a 1.7 µm column for the separation of a six-compound mixture, (b) same method condition except the column is packed with 5 µm particles, and (c) the same 5 µm column as in (b), and run with the same method conditions except the adjustment done in ABPR pressure to 3,390 psi in place of 1,500 psi. Reprinted with permission from Ref. [8].

3.5 Conclusion

The main difference between an (U)HPLC system and a packed-column SFC system is the deployment of CO_2 as a solvent in the latter. CO_2 is significantly more compressible compared to solvents used in HPLC, which challenges robust operation of individual system components and also the system as a whole. In addition, absence of water in SFC facilitates reactions that are not commonly observed during (U)HPLC operations. All these factors together made SFC operation with old instrumentation and columns an untrustworthy endeavor.

Instrument and column research conducted over a period of time could clearly identify the causes that bring variability in SFC operation. This chapter presents discussions on each of these topics, explaining the problem and how solutions were developed. The chapter demonstrates that state-of-the-art SFC instruments and SFC-specific columns are designed to mitigate any unwanted variability, which should make SFC as trustworthy as any (U)HPLC instrument. A recent study (see ref. [10]) on method reproducibility with a commercial SFC instrument confirms this claim. The study involved 19 participating laboratories from 9 different countries across 4 continents. The participants included 5 academic groups, 3 demonstration laboratories at analytical instrument companies, 10 pharmaceutical companies and 1 food company. Results demonstrated repeatability and reproducibility variances either similar or better than the ones described for LC methods.

Acknowledgments

The authors acknowledge Jacob Fairchild of Waters Corporation for supplying the original figures that were published in Refs. [3, 6] and also for valuable discussion. They are thankful to Thomas Swann of Waters Corporation for critical comments and fruitful discussion and are also thankful to Carsten Wess of Waters Corporation for his work generating Figure 3.6.

References

[1] Online refractive index database, https://refractiveindex.info, accessed on 6 October 2017.
[2] G. J. Besserer, D. B. Robinson, Refractive indices of ethane, carbon dioxide, and isobutane, J. Chem. Engg. Data, 18 (1973) 137–140.
[3] J. N. Fairchild, D. W. Brousmiche, J. F. Hill, M. F. Morris, C. B. Boissel, K. D. Wyndham, Chromatographic evidence of Silyl Ether Formation (SEF) in supercritical fluid chromatography, Anal. Chem. 87 (2015) 1735–1742.
[4] K. Ebinger, H. N. Weller, Comparative assessment of achiral stationary phases for high throughput analysis in supercritical fluid chromatography, J. Chromatogr. A 1332 (2014) 73–81.
[5] K. D. Wyndham, J. E. O'Gara, T. H. Walter, K. H. Glose, N. L. Lawrence, B. A. Alden, G. S. Izzo, C. J. Hudalla, P. C. Iraneta, Characterization and evaluation of C18 HPLC stationary phases based on ethyl-bridged hybrid organic/inorganic particles, Anal. Chem. 75 (2003) 6781–6788
[6] J. N. Fairchild, D. W. Brousmiche, Explaining Silyl Ether Formation (SEF) in supercritical fluid chromatography (SFC), LCGC, 11 (2015) 12
[7] J. F. Hill, A. Tarafder, E. Chambers, Considering mobile phase compressibility to develop a Robust method in SFC, Waters Corporation Application Notes: 720005548EN Author.
[8] A. Tarafder, C. Hudalla, P. Iraneta, K. J. Fountain, A scaling rule in supercritical fluid chromatography. I. Theory for isocratic systems, J Chromatogr. A, 1362 (2014) 278–293
[9] A. Tarafder, Jason F. Hill, Scaling rule in SFC. II. A practical rule for isocratic systems, J Chromatogr. A, 1482 (2017) 65–75.
[10] Amandine Dispas et al, First inter-laboratory study of a Supercritical Fluid Chromatography method for the determination of pharmaceutical impurities, DOI: https://doi.org/10.1016/j.jpba.2018.08.042

Craig White, Andreas Kaerner, Alfonso Rivera, Eric Seest, Matthew Belvo, John Burnett, Thomas Perun, Arancha Sonia Marin, Thomas Castle, Maria Luz de la Puente, Cristina Anta, Pilar Lopez

4 SFC within early drug discovery research – a user's perspective

4.1 Introduction

Supercritical fluid chromatography (SFC) was first introduced to Lilly Research Laboratories in 2003 as a complimentary tool to high-performance liquid chromatography (HPLC) for chiral analysis and purification. Nowadays, and 14 years later, SFC is embedded as the primary separation and purification tool (for chiral) across all three research sites. The speed, separation orthogonality to traditional chromatography approaches, and solvent handling benefits of SFC are fully realized from early drug discovery (lead generation), to large-scale and late-phase preclinical toxicology. This chapter will highlight the learning and practical experiences from the SFC end user with an emphasis on applications relevant to the medicinal chemist. Topics include (i) the challenges of supplying CO_2 to laboratory instrumentation, (ii) chiral screening and scale-up, (iii) the benefits and limitations of mobile-phase additives, (iv) automating large-scale purification, and (v) high-throughput achiral purification.

4.2 CO_2 infrastructure

One of the first questions to consider when implementing SFC technology in the laboratory is how to deliver CO_2 to instrumentation. In practice, there are several delivery systems available for supplying pressurized CO_2. Examples include:
- Small-volume cylinders where the liquid CO_2 is pulled from the base of the cylinder using a dip tube at an appropriate pressure.
- Medium-volume dewars where the liquid CO_2 is removed using a dip tube and recompressed to a higher working pressure.
- Large-volume bulk tank systems where the gas is extracted from the top of the liquid tank and compressed to liquid using an appropriate booster pump.

Craig White, Eli Lilly & Company, United Kingdom
Andreas Kaerner, Eli Lilly & Company, USA
Alfonso Rivera, Maria Luz de la Puente, Cristina Anta, Pilar Lopez, Eli Lilly & Company, Spain
Eric Seest, Matthew Belvo, Thomas Castle, Eli Lilly & Company, USA
John Burnett Eli Lilly & Company, United Kingdom
Thomas Perun, Arancha Sonia Marin, Eli Lilly & Company, Spain

https://doi.org/10.1515/9783110500776-004

The benefits and limitations of each approach are summarized in Table 4.1.

Table 4.1: Benefits and Limitations of typical CO_2 delivery systems.

	Benefits	Limitations
Cylinders	– Simple installation – Single cylinder hookup adequate for analytical SFC instrumentation	– Battery of cylinders required to support higher flow rates – Significant overhead to swap cylinders for continuous work – Independent measurement systems required to monitor CO_2 level, e.g., weighing station – High running cost compared to alternative solutions
Dewars	– Provides greater capacity for purification systems – Automated switching devices can provide uninterrupted operation, e.g., overnight operation	– Liquid or gas booster pump required – Facilities management required to regularly replace empty dewars – Safety risk assessments required for installations made inside the work area – Heaters and appropriate ventilation required – Premature switching when the dewar environment is cold or when CO_2 demand is high
Bulk tank	– Unlimited capacity – Continuous and uninterrupted operation across the working week – Inexpensive industrial grade CO_2. High purity gas is drawn from the top of the tank.	– Installation of external CO_2 tank requires significant up-front investment – Gas booster pump required

For the installation of an analytical SFC instrument, a simple CO_2 cylinder setup close to the instrument is adequate. However, when operating multiple instruments for analytical analysis and/or method development screening as well as SFC purification instruments, the total consumption could reach the level of 1 kg of CO_2 per minute or more. This requires the installation of a bank of cylinders with a second set as backup, with automated pressure switching when the temperature and pressure fall on the primary set during heavy use.

For both the dewar and bulk tank options, a pressurization system is required to deliver high-pressure liquid CO_2 to the SFC instrument. There are two formats: (i) a pressurizing pump for liquid CO_2 and (ii) a booster pump for CO_2 gas. The benefits and limitations of each are described in Table 4.2.

Table 4.2: Benefits and Limitations of CO_2 compression systems.

	Benefits	Limitations
Liquid pump	– Low cost and maintenance	– Requires high purity liquid CO_2 before feeding the delivery system from a cylinder dip tube – When using dewars, the pump may freeze after continued heavy use requiring a temporary shutdown
Gas pump	– Industrial quality liquid CO_2 can be used – The booster pump operates according to the CO_2 demand from the lab	– Increased maintenance overhead – Spare booster pump recommended as a breakdown contingency

At Lilly, we are working with different configurations depending on needs and regulations. In some installations, air-driven booster pumps are used to compress gaseous CO_2 supplied from bulk tanks. Two compression steps elevate the supply gas pressure from 200 to 1,500 psi. Preboosters first pressurize the CO_2 from 200 to 700 psi and then the main booster pump completes the compression step to 1,500 psi. This two-step compression process enables the delivery of up to a maximum of 2 kg of CO_2 per minute. This is the sufficient capacity to meet the CO_2 demand of most analytical and preparative SFC laboratories.

Another option is the use of dewar delivery. To meet a moderate CO_2 demand, a system that draws liquid from two 180 L dewars (up to 500 g/min) is adequate. To meet a high CO_2 demand, where staging a bulk CO_2 tank is not an option, some labs combine (i) a system that recompresses gas from the top of two banks containing two 180 L dewars each and (ii) a CO_2 bulk liquid pumping system that draws liquid CO_2 from two banks of three 230 L dewars each. The total combined capacity is approximately 1 kg/min as staged.

The CO_2 delivery system sits on the critical path for all SFC analytical and purification labs. Robustness and reliability are therefore key to success. As described there are multiple flavors of CO_2 delivery systems in use across Lilly; strategically we favor the bulk tank design as the deployment of SFC increases.

4.3 Chiral method development and scale-up

At Lilly, the area to benefit most from SFC technology is chiral analysis and purification. The impact has been dramatic resulting in a significant shift from HPLC to SFC technology. This section will focus on our experience with chiral method development with two goals in mind: (i) for chiral purity determination and (ii) for preparative scale purification.

The goal of method development is to find a suitable separation by rapidly screening a large matrix of columns and solvents. Generally, the primary screen consists of running fast gradients on samples across 12 different methods using methanol (MeOH), ethanol (EtOH), and isopropanol (IPA) as cosolvents, each with a basic additive such as 0.2% isopropylamine (IPAm), and four traditional Daicel columns, for example, Chiralpak AD-H, Chiralpak AS-H, Chiralcel OD-H, and Chiralcel OJ-H. If a satisfactory separation is not achieved, a second tier of columns is explored such as Phenomenex Lux (Cellulose-2, Cellulose-4, and Amylose-2), Daicel (Chiralpak IG, Chiralpak IC, Chiralpak IA, and Chiralpak IB), and Regis Whelk-O1 stationary phases using gradient conditions. This tiered approach has led to a ~98% success rate for adequate separation of stereoisomeric mixtures for subsequent preparative scale-up.

The chiral screening methodology has changed over time to incorporate improvements in column technology and instrumentation. Improvements have been made by implementing stationary phases with smaller particle sizes (shifting from 10 to 5 to 3 µm), moving to smaller columns (from 4.6 × 250 to 4.6 × 100 to 3.0 × 100 mm), along with shorter run times (reducing from 10 to 5 to 2 min) [1–3]. The use of column switching, solvent selection, and software to automate the multiple experimental combinations has also helped to speed up the process [4, 5].

In many cases, the complete set of experiments does not need to be completed. For methods to determine enantiomeric excess or diastereomeric ratio, no further method development is required when a suitable separation is found. Likewise for preparative work, the method development screen can be truncated if a hit is achieved. However, from a practical perspective this occurs less frequently due to the more stringent resolution and run time criteria employed for purification method scouting.

SFC has an advantage over HPLC because the mobile phase is run at a flow rate that is five times greater (for the same dimension column) than HPLC. This is due to the lower viscosity and higher diffusivity of mobile phases containing CO_2. Better cosolvent miscibility provides another advantage of SFC, allowing for the use of three cosolvents: MeOH, EtOH, and IPA instead of just EtOH and IPA for heptane-containing normal-phase HPLC methods. Screening with all these solvents is essential since many times only one of the cosolvents provides separation. In addition to screening single columns in sequence, we have instruments capable of screening five columns in parallel [6].

As with HPLC screening, amine additives are used in the mobile phase to significantly reduce the tailing of basic molecules (this is discussed in more detail in the next section of this chapter). However, unlike HPLC, the use of acidic additives is generally not needed to chromatograph compounds containing carboxylic acids, tetrazoles, or phenols, due to the acidic nature of CO_2. The lack of acidic additives avoids the formation of undesired esters during the workup process which is prevalent when separations are performed using reversed-phase (RP)-HPLC with MeOH.

Following the completion of the analytical screen, the gradient method is converted to an isocratic method for preparative separations using "stacked injections" or for chiral purity analysis. Stacked injections are a more efficient way to process material since multiple separations are performed on the chromatography column without a column re-equilibration step. Injections are timed to allow the second group of peaks not to overlap the first, the third group not to overlap the second, and so on. Generally, the second injection is made before the first group has eluted from the column. Injection to injection cycle times are usually from 2 to 5 min. Preparative separations are performed using the same mobile phase as the analytical isocratic method at flow rates of 65–80 mL/min for 21.2 mm ID columns, 150–200 mL/min for 30 mm ID columns, and 250–400 mL/min for 50 mm ID columns – all containing 5 μm stationary phase. Scale-up is almost always successful with mixture loadings of between 5 and 200 mg for a 21.2 mm column. For more difficult separations, 250 mm length columns are used, but for the majority of separations a 150 mm length column is sufficient [7]. Early on it became clear that SFC was not only faster for method development, compared to HPLC, but also provided separations not achievable by any other means. Therefore, HPLC method development is rarely required, and today most preparative chiral separations are performed using SFC [8–10].

4.4 Additives for SFC

Mobile-phase additives used for chiral and achiral SFC can be crucial to obtain acceptable chromatographic performance [11]. Such additives function by buffering the pH and masking unwanted analyte–stationary phase interactions. For example, the use of amine additives greatly improves the poor peak shape often observed for basic compounds by suppressing interaction of the analyte with uncapped silanol groups on the stationary phase. In our experience, basic additives are essential but introduce their own challenges to the chromatographic process. Issues with removal after purification, analytical to preparative scalability, odor, solubility, and column stability have all been observed in our laboratories. The additive of choice for normal-phase liquid chromatography, 0.2% IPAm, was employed at Lilly for SFC with excellent chromatographic results across both analytical and preparative scales. However, nuclear magnetic resonance identified many isolated fractions containing residual IPAm which was not fully removed during evaporation. The isopropylcarbamate salt of the target of interest (when formed) can be difficult to remove and accounts for higher than expected weight recovery from collected fractions. Ammonia (20 mM) in MeOH is an alternative additive for preparative SFC that affords sharp peaks and improved mass spectrometry (MS) ionization, and the risk of forming salts postpurification is eliminated. However, the reaction of carbon dioxide with ammonia to form ammonium carbamate quickly occurs when the CO_2 and modifier streams meet. Ammonium carbamate is soluble in MeOH and EtOH, but for IPA, a specific blend of IPA: MeOH (99:1) is required to maintain solubility. Careful consideration is still required before using ammonia as the basic additive for purification. Certain chiral columns are not tolerant to ammonia and certain instrument pumping configurations facilitate the precipitation of ammonium carbamate within the pump head when the mixture of CO_2 and IPA is cooled to $-5\,°C$.

Tertiary amine additives (triethylamine, diethylmethylamine [DEMA], and dimethylethylamine [DMEA]) deliver acceptable chromatographic results but their ultraviolet (UV) absorbance at low wavelengths greatly reduces the linear UV dynamic range and inhibits accurate analytical work below certain wavelengths. This is usually not an issue for preparative purifications since they don't rely on a large linear dynamic UV range. Due to high loading, the chromatographic peak UV absorptions are already outside the linear range of the detector, or have sufficient signal at longer wavelengths. DMEA is generally the additive of choice for preparative purification as it is the most volatile tertiary amine and is therefore the easiest additive to remove during dry down. IPAm is the additive of choice for analytical work due to its UV transparency down to about 200 nm and can also provide suitable chromatography conditions that typically scale up well to preparative purification.

4.5 Walk-up chiral purification

The demand for chirally pure compounds within early drug discovery has increased significantly during the last 10 years, heavily impacting the chiral purification workload and increasing the need for innovative approaches to reduce overall cycle time. Efforts to automate analytical chiral screening have been discussed earlier in this chapter. This section will highlight the opportunity to balance workload between the medicinal chemist and the chromatography expert by offering chiral purification as a walk-up or open-access (OA) service.

The chiral purification workflow can be categorized into two streams: (i) submissions requiring the involvement of skilled personnel due to high complexity, for example, tight separation, the presence of multiple chiral centers, and/or solubility issues; and (ii) noncomplex requests, for example, small quantities of material, facile separation, and/or when chemistry teams require material with lower optical purity for biological testing. Medicinal chemists have long depended on OA RP-HPLC-MS because the user-friendly and reliable instrumentation rapidly produces high-quality results. As a result, both analytical and preparative achiral HPLC-MS systems are routinely found in conventional chemistry laboratories to support OA activities. Unfortunately, a similar approach for noncomplex "stream 2" chiral SFC purification is challenged by the lack of such user-friendly, robust, and reliable instrumentation. However, recent technological advances have provided an opportunity to collaborate with instrument providers with the goal to establish a fully automated analytical screening and purification SFC-UV platform for OA chiral purification.

Software enhancements have enabled automation of the purification workflow such as: (i) the identification of target isomers through UV spectra matching, (ii) the selection of conditions to simultaneously achieve chiral discrimination and achiral specificity, (iii) the translation of gradient screening conditions into isocratic elution for purification, and/or (iv) the definition of time windows for efficient stacked injection and precise collection of isomers.

Figure 4.1 shows an example of the results achieved using such a fully automated process. Analytical gradient screening is followed by automated data analysis to identify the best resolving conditions based on a decision tree and scoring criteria (Figure 4.1a). This information is used to create a preparative method with isocratic elution conditions and UV collection parameters. The resulting chromatogram from a pilot injection determines the key parameters for efficient scale-up, for example, stacked injection and collection windows definition (Figure 4.1b). Results from this

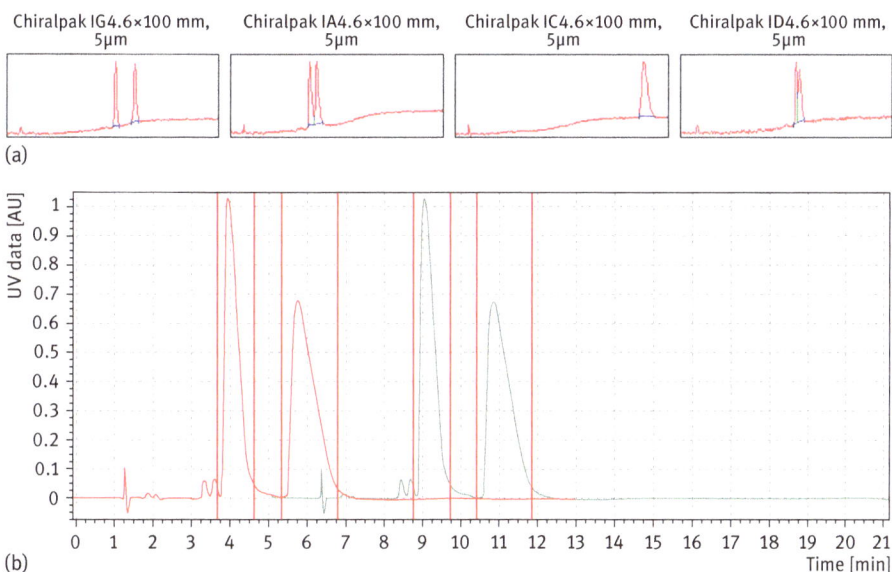

Figure 4.1: (a) The IG column delivers the best separation with a resolution factor of 4.32, compared to 1.38 (IA column), no separation (IC column) and 0.53 (ID column); (b) the red trace shows the elution profile from the scouting run, the green trace is the virtual representation of the stacked injection, calculated by the software, to define the time windows for injection; (c) purification campaign; and (d) final purity assessment.

Note: Experimental conditions: (a) Screen: Chiralpak IG (5 µm, 4.6 × 10 mm); A:CO_2/B:MeOH (0.2% DMEA); gradient from 15% to 55% in 2 min, then 2 min at 55%; 5 mL/min, 120 bar, 40 °C. (b) and (c) Preparative: Chiralpak IG (5 µm, 2 × 25 cm); 6/4 CO_2/MeOH (0.2% DMEA); 65 mL/min, 100 bar, 40 °C; 25 mg every 5.1 min. (d) Optical purity assessment: Chiralpak IG (5 µm, 4.6 × 10 mm); A:CO_2/B:MeOH (0.2% DMEA); gradient from 15% to 55% in 1.5 min, then 1 min at 55%; 4 mL/min, 120 bar, 40 C

(c)

(d)

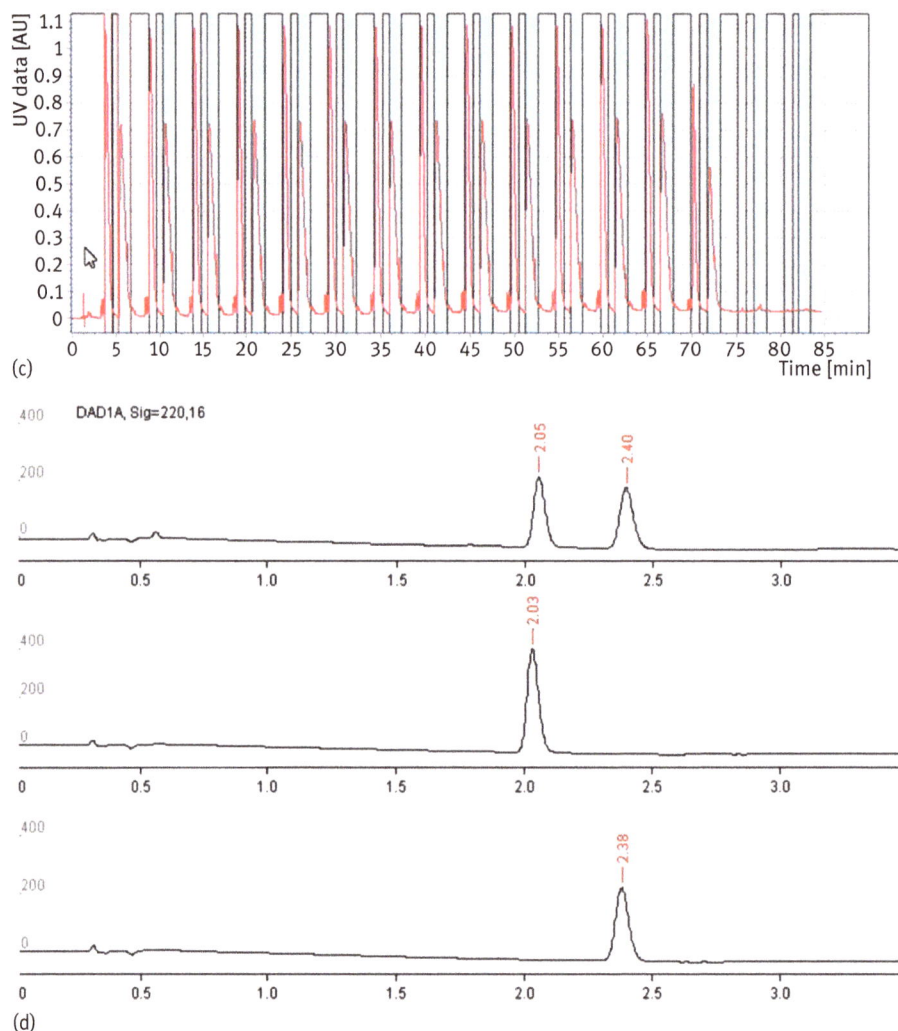

Figure 4.1: (continued)

scouting run are used as reference to start the purification campaign (Figure 4.1c). The software continuously monitors peak UV height to determine run completion triggering a subsequent automated wash protocol to ready the system for the next sample. Purity assessment of isolated fractions is carried out using alternative analytical OA-SFC-MS units equipped with the same set of stationary phases (Figure 4.1d). The OA user interface provides access to chemists (or analysts) to simplify the submission process. The user walks up to the instrument, enters a unique identifier, and submits two vials – one for the analytical screen and the second for the purification injections with each prepared at specific concentrations. Generally pure isomers can be collected within 2 h of submission.

This OA approach can benefit both the medicinal chemist and the experienced analyst to quickly process small-scale samples during normal working hours and to support larger scale samples overnight. In our experience, this highly efficient and simple to use OA SFC platform can successfully streamline the workflow for those "stream 2" chiral purifications, enabling experts to focus on more challenging samples, method development, and research projects.

4.6 Automating large-scale purification

As discussed, SFC is the preferred choice for preparative enantioseparation within drug discovery where semipreparative SFC (3 cm i.d. column size) is routinely employed to purify intermediates and final products in quantities up to 50 g. As compounds progress through the medicinal chemistry pipeline, greater quantities of material are required; this can be in the range of 200 g of pure enantiomer for those compounds requiring toxicological evaluations. To meet the increasing demand for larger scale purification (achiral and chiral) within discovery, and to deliver in a timely manner, SFC is utilized in an uninterrupted batch process where purification campaigns are run unattended overnight across multiple days.

Unlike a manufacturing environment, handling large volumes of organic solvent within a discovery laboratory environment can be problematic due to fire safety regulations that impose tighter limitations on flammable solvent utilization. With such solvent restrictions in mind, SFC purification campaigns are advantageous over HPLC as they generally require 5–40% (v/v) organic modifier, with the majority of the mobile phase (CO_2) expanding to gas, and providing an added benefit of reduced evaporation time. Although SFC significantly decreases organic solvent consumption, automating unattended overnight operation does require careful planning and changes to lab infrastructure. The risks associated with running at a flow rate up to 400 g/min require the appropriate health and safety controls such as classification of areas into hazardous zones, solvent and CO_2 detectors, external alarms, air-driven pumps for solvent transfer; and automated shutdown procedures with valves to isolate the modifier and CO_2 solvent supply from the purification instrument. Managing the CO_2 supply from a bulk tank delivery system is highly advantageous, and when combined with CO_2 recycling technology, the CO_2 consumption and cost is dramatically reduced. The bulk tank also makes the process scalable, enabling expansion to multiple systems working in parallel. The solvent evaporation process then becomes the bottleneck where fraction collection and automated evaporation need to be concurrent.

Instrument reliability and chromatographic reproducibility are critical to a successful multiday purification campaign, especially when there are no manual intervention opportunities overnight. Figure 4.2 illustrates the potential of SFC to purify large amounts of material when used in automation overnight. For this example,

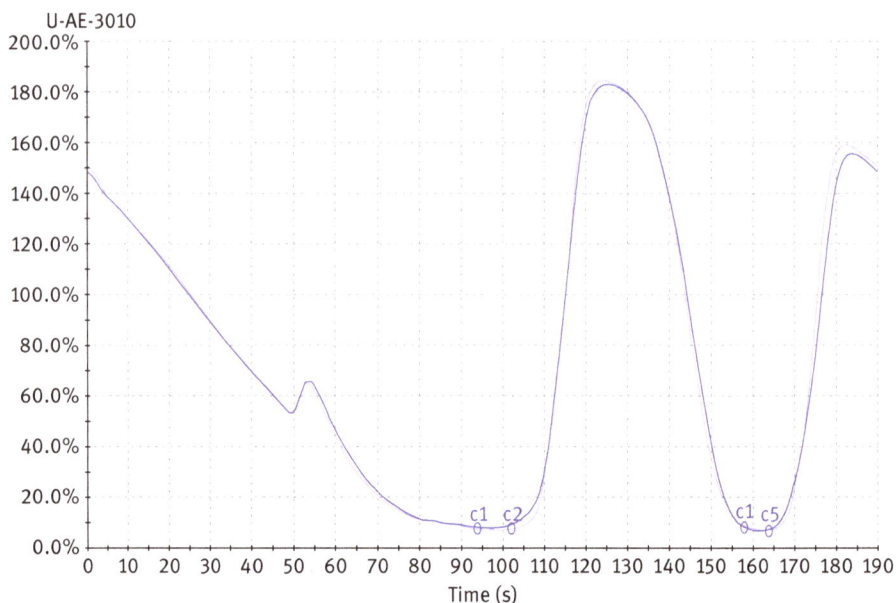

Figure 4.2: Preparative chiral SFC example (902 g) showing two overlaid chromatograms from a run of 820 injections across 2 days. The flow rate was 350 g/min using 22% ethanol with diethylmethylamine (0.2%) using an AD-H column of dimensions 5 cm i.d. × 25 cm (5 µm). Injection amount was 1.1 g (20.8 g/h). The first eluting isomer was collected into position 2 and the second eluting isomer collected into position 5.

902 g of material was processed, and pure enantiomers returned, within 3 working days of submission. Approximately 1.1 g of racemate was injected every 190 s using only a total of 260 L of organic solvent. In our experience, SFC is generally a faster and more cost-effective option compared to HPLC.

Efficient chiral SFC purification requires adequate solubility of the sample in MeOH, EtOH, or IPA (>20 mg/mL preferred). More traditional coated polysaccharide-based Chiral Stationary Phase (CSP) can tolerate small amounts of alternative solvents as an additive to the sample diluent when solubility is an issue. Chloroform is preferred in our experience and is used routinely in amounts up to 1 mL/125 g of CSP (Dichloromethane (DCM) to a lesser degree). Immobilized versions of the polysaccharide chiral stationary phase can also be employed with a wider range of cosolvents and sample diluents to improve sample solubility with less risk of phase bleeding. Incremental loading should be employed to determine whether sample precipitation might occur when the sample band meets the modifier/CO_2 stream. HPLC is still employed to a lesser degree for chiral separation in cases where more polar mobile-phase conditions are helpful and where resolution and campaign metrics are favorable when compared to SFC.

The application of SFC for overnight, unattended purification offers significant benefits over HPLC for discovery research purification laboratories. The reduced organic solvent consumption and fast chromatography changes the purification landscape, enabling purification work which would otherwise not be considered due to solvent-handling complications. Without the infrastructure to support automated large-scale SFC purification, 50 g is a realistic glass ceiling for daytime SFC operation using 3 cm i.d. hardware. By utilizing nighttime capacity for single-column (5 cm i.d.) batch preparative SFC, purifications up to 1 kg are possible. A 1-week cycle time is a reality for samples in the region of 500 g. This capability is highly valued for those compounds synthesized at scale for toxicological evaluations, where speed is the key driver.

4.7 Automated achiral SFC purification

Following the successful application of SFC for chiral analysis and purification, the potential of SFC to support achiral purification has been explored with the goal to complement the well-established RP-HPLC-MS approach used in our high-throughput purification (HTP) laboratories.

A key element in any purification strategy is identification of a generic set of analytical columns that provide sufficient orthogonality to successfully resolve analytes from their respective impurities across the diverse chemical landscape encountered in discovery. Initial efforts using 2-ethylpyridine (2-EP) and cyano phases found that they did not provide sufficient orthogonality to chromatographically resolve the mixtures entering the purification workflow. Subsequent efforts resulted in a five-column strategy consisting of 2-EP, benzenosulfonamide, diethylaminopropyl, diol and dinitrophenyl chemistries which were selected based on their broad applicability and complementary properties. It was found that more than 85% of mixtures can be separated using this five-column strategy [12]. In cases where serial screening was implemented, further improvements in stationary-phase development allowed for a decrease from five to two columns. They consisted of a hydrophilic interaction chromatography cross-linked diol chemistry and original 2-EP, while maintaining the aforementioned separation success rate [13]. In addition to UV detection, MS was used to reliably identify the analyte of interest.

With the establishment of a robust analytical screening capability, the focus can now be turned to scale up strategy that yields a similar level of success. In general, the analytical column type and conditions that afforded the best chromatographic separation were used to select the preparative column type and associated generic conditions. While it is well established that chromatographic behavior can also be manipulated by changing temperature, pressure, and mobile-phase additives, it was determined that this level of chromatographic customization on the analytical and preparative scales was not well suited for a high-throughput lab and instead left for the exceptions.

The progress of SFC for HTP was hindered by (i) the absence of the universal achiral SFC column, (ii) poor instrument reliability, and (iii) poor instrument robustness. Mixtures submitted for purification were typically screened across multiple columns to determine the best chemistry for scale-up; additional time was therefore required for method development compared to the more generic HPLC approach. High-throughput SFC purification was limited to those cases where RP-HPLC screening failed to separate the component of interest from impurities, or when RP-HPLC purification failed to deliver final purity requirements due to unexpected scale-up issues. Figure 4.3 shows an example of the first scenario. Purification workflows are now designed to accommodate both HPLC and SFC techniques with the incorporation of bespoke options to improve the probability of success when

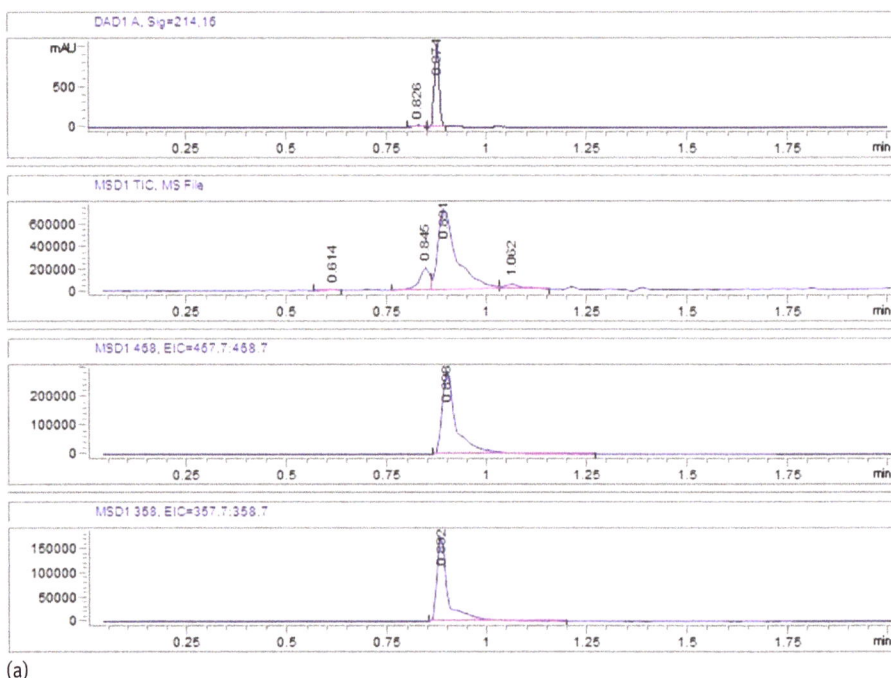

(a)

Figure 4.3: (a) All conditions in routine RP-HPLC-MS screen showed coelution of the main impurity (MW 358) with the compound of interest (MW 468). Chromatogram in (a) shows the results using the XBridge C18 (3.5 μm, 2.1 × 50 mm); A: H_2O – 10 mM NH_4HCO_3 pH 9/B: acetonitrile; gradient from 10% to 100% B in 1.5 min, then hold for 0.5 min at 100% B; 1.2 mL/min, 50 °C. (b) SFC generic gradient screening. 2-Ethylpyridine (2-EP) and Luna Hilic (Hilic diol), both 5 μm, 4.6 × 150 mm; A: CO_2; B: MeOH – 10 mM NH_4HCO_3; gradient from 10% to 40% B in 1.7 min and then hold for 0.5 min at 40% B; 120 bar; 4 mL/min, 40 °C. (c) SFC purification where (a) is the UV chromatogram at 215 nm showing the DMSO peak at 1.03 min, the impurity at 1.74 min, and the compound of interest at 3.57 min; (b) the single ion mass chromatogram for the target of interest (ESI positive mode). Conditions: 2-EP (5 μm, 30 × 150 mm); A:CO_2/B:MeOH–10 mM NH_4HCO_3; 1 min at 20% then gradient from 20% to 40% B in 3 min and then hold for 1 min at 40% B; 120 bar; 100 mL/min, 40 °C.

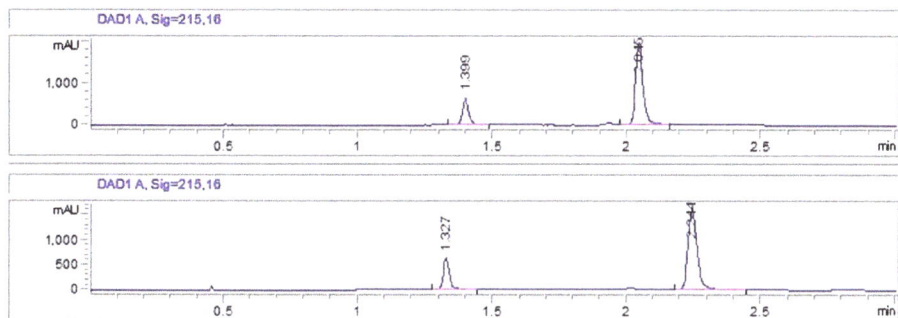

(b)

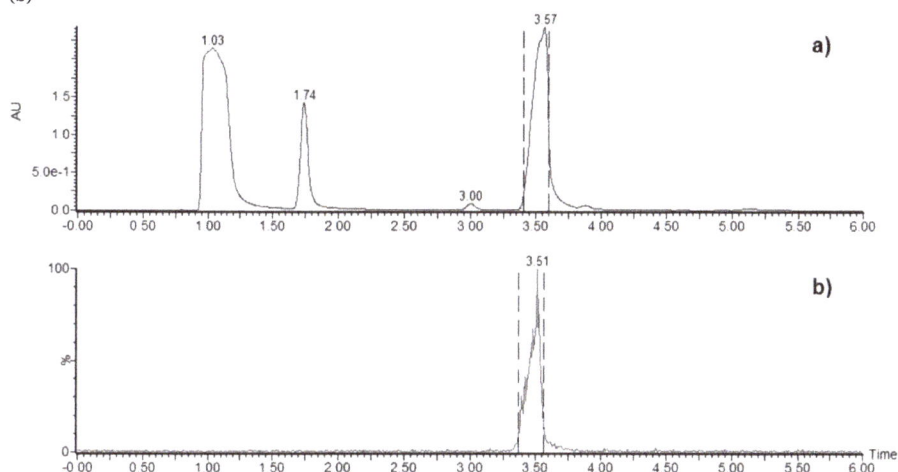

(c)

Figure 4.3: (continued)

using SFC. For example, we experienced frequent reliability and robustness issues when SFC was directly applied to the purification of crude mixtures from automated synthesis platforms [14]. In this environment, compounds to be isolated are often a first attempt at synthesis that have been prepared at the direction of a remote user that has little first-hand knowledge of the reaction mixtures' solubility characteristics, composition, or physical appearance when submitted to the purification labs. It is our experience that RP-HPLC-MS is much more robust and forgiving than SFC for chromatographically complex mixtures containing a broad range of polar and nonpolar components. To overcome the problems caused by widely varying solubilities from components in complex mixtures, implementation of a solid-phase extraction sample pretreatment step resulted in improved reliability.

However, even with an adequate cleanup procedure for complex mixtures, solubility in liquid CO_2 is still not easily predictable and many factors affect this property [15]. A few strategies have been reported to be employed to ascertain solubility information as (i) models based on analytical retention times [16], linear solvation energy [17], Flory–Higgins equation [18], or sublimation enthalpies [19] or (ii) experimental determinations using supercritical fluid extraction cells [20, 21] or continuous flow apparatus [22]. Neither has yielded results that could be used in preparation of SFC purification, nor has been, to the best of our knowledge, successfully incorporated into workflows in any analytical laboratories dedicated to the purification task in the pharmaceutical industry. Furthermore, some authors have already reported that results can vary when using different solubility methods, also highlighting the importance of data interpretation [23].

To expand the applicability of SFC to broader structural classes (e.g., metabolites, peptides, and peptidomimetics), the addition of water to the mobile phase has been explored [24]. The addition of water (i) extends the polarity and strength of cosolvents, (ii) improves solubility of polar analytes, and (iii) helps retain water in the stationary phase providing a Hydrophilic Interaction Chromatography (HILIC)-type interaction between analytes and stationary phase. Separations of mixtures and ten drug-like analytes exhibited superior selectivity and excellent peak resolution when carried on bare silica columns with the addition of water to MeOH or to three different additives (trifluoroacetic acid, isopropyl amine, and ammonium acetate) [25].

There is no doubt that one of the most challenging problems encountered during the development of a new pharmaceutical compound is linked to purification. From this perspective, analytical laboratories in pharmaceutical companies focus on building technology platforms that integrate processes, technologies, methodologies, and data management. The leading technology involved in this scenario seems to still be RP-HPLC-MS. Nevertheless, the availability of dual automated operating platforms is nothing but good for our research business. Furthermore, with the hope that new technology at preparative scale (MS guided) becomes a reality, we would like to foresee a new scenario for achiral SFC purification in the OA environment that sits alongside low-pressure Flash and HPLC as primary separation techniques available to research chemists. We have not established a process to evaluate CO_2 solubility based on any of the reported strategies yet, but are cognizant of the importance in doing so in order to reliably process crude reaction product mixtures in a high-throughput manner. Fortunately, recent developments have enabled the possibility of using SFC in routine achiral purification: (i) an increase in the breadth of available column chemistries and column vendors, (ii) the promising results observed with the addition of water to mobile phase; or (iii) more important, although still mostly at analytical scale, the improved performance of instrument hardware and software that is capable of withstanding the rigors of OA utilization.

Acknowledgments

The authors thank the publishers of *Chromatography Today* for their permission to reproduce sections of the article in *Chromatography Today* May/June 2015.

References

[1] C. Hamman, M. Wong, M. Hayes, P. Gibbons, J. Chromatogr. A 2011, 1218, pp.3529–3536.

[2] C. Hamman, M. Wong, I. Aliagas, D.F. Ortwine, J. Pease, D. E. Schmidt Jr., J. Victorino, J. Chromatogr. A 2013, 1305, pp. 310–319.

[3] E.L. Regalado, Christopher J. Welch, J. Sep. Sci. 2015, 38, pp. 2826–2832.

[4] W. Schafer, T. Chandrasekaran, Z. Pirzada, C. Zhang, X. Gong, M. Biba, E. I. Regalado, C. J. Welch, Chirality, 2013, 25, pp. 799–804.

[5] D.C. Patel, M.F. Wahab, D.W. Armstrong, Z.S. Breitbach, J. Chromatogr. A, 2016, 1467, pp. 2–18.

[6] L. Zeng, R. Xu, D.B. Laskar, D.B. Kassel, J. Chromatogr. A, 2007, 1169, pp. 193–204.

[7] D. Speybrouck, E. Lipka, J. Chromatogr. A, 2016, 1467, pp. 33–55.

[8] N. Wu, Advances in Chromatography, 2008, 46, pp. 213–234.

[9] D. Speybrouck, E. Lipka, J. Chromatogr. A, 2016, 1467, pp. 33–55.

[10] H. Leek, L. Thunberg, A. C. Jonson, K Ohlen, M. Klarqvist, Drug Discovery Today, 2017, 22 (1), pp. 133–139.

[11] E. Lesellier, C. West, J. Chromatogr. A, 2015, 1382, pp. 2–46.

[12] M.L. de la Puente, P. López-Soto and J. Burnett, J. Chromatogr. A 2011, 1218, 47, 8551–8560.

[13] M.L. de la Puente, P. López-Soto, C. Anta, J. Chromatogr. A 2012, 1250, 172–181.

[14] A.G. Godfrey, T. Masquelin, H. Hemmerle, Drug Discovery Today, 2013, 18, Numbers 17/18, pp. 795–802.

[15] E. Lesellier, J. Chromatogr. A, 2009, 1216, 10, pp. 1881–1890.

[16] J. Yang, P.R. Griffiths, Anal. Chem., 1996, 68, 14, pp. 2353–2360.

[17] N. De Zordi, I. Kikic, M. Moneghini, D. Solinas, J. Supercrit. Fluids, 2012, 66, pp. 16–22.

[18] C. Su, Y. Chen, Fluid Phase Equilibria, 2007, 254, pp. 167–173.

[19] A. Tabernero, E.M. Martín del Valle, M.A. Galán, Ind. Eng. Chem. Res. 2013, 52 (51), pp. 18447–18457.

[20] Y. Yamini, J. Hassan, S. Haghgo, J. Chem. Eng. Data (2001), 46, pp. 451–455.

[21] K.H. Gahm, K. Huang, W.W. Barnhart, W. Goetzinger, Chirality, 2011, 23, 1E, pp. E65–E73.

[22] G.-T. Liu, K. Nagahama, J. Supercrit. Fluids, 1996, 9, Issue 3, pp. 152–160.

[23] I. Rodríguez-Meizoso, O. Werner, C. Quan, Z. Knez, C. Turner, J. Supercrit. Fluids, 2012, 61, pp. 25–32.

[24] L.T. Taylor, M. Ashraf-Khorassani, J. Sep. Sci., 2010, Vol33, 11, pp. 1682–1691.

[25] L.T. Taylor, M. Ashraf-Khorassani, E.T. Seest, J. Chromatogr. A, 2012, 1229 pp. 237–248.

Vincent Desfontaine, Raul Nicoli, Tiia Kuuranne,
Jean-Luc Veuthey, Davy Guillarme

5 What is the potential of SFC-MS for doping control analysis?

Abstract: This chapter describes the application of modern supercritical fluid chromatography (SFC), coupled with mass spectrometry (MS), for the analysis of various classes of doping agents. Even if liquid chromatography (LC) and gas chromatography (GC) remain the reference techniques for doping control analysis, SFC appears as a promising strategy for the determination of numerous doping agents owing to the recent progresses. Indeed, this chromatographic technique appears as a faster, greener, more universal, and orthogonal analytical method compared to LC-MS and GC-MS. It also offers improved sensitivity, lower matrix effects, and excellent precision when analyzing biological fluids. For all these reasons, SFC-MS is expected to be more and more widely used in doping control laboratories.

Keywords: supercritical fluid chromatography, mass spectrometry, doping control analysis, prohibited substances

5.1 Introduction to doping control analysis

Since 1999, the rules for fighting against doping in sports are established by the World Anti-Doping Agency (WADA), an independent institution that has implemented the *World Anti-Doping Code*. The objective of this Code is to harmonize anti-doping policies, rules, and regulations within sport organizations, and to provide regulations for the detection, deterrence, and prevention of doping [1]. To ensure worldwide harmonization, there are currently five *International Standards*, covering different technical aspects of the Code: the *List of Prohibited Substances and Methods* (List), *Testing and Investigations*, *Laboratories*, *Therapeutic Use Exemptions* (TUE), and *Protection of Privacy and Personal Information*. All anti-doping laboratories should be compliant with the prevailing regulations and should be accredited by both ISO17025 and WADA for performing urine and blood sample analysis [2]. There are also some additional technical documents describing deeper details of specific processes, for example, the identification criteria for gas chromatography (GC) and liquid chromatography (LC) coupled to mass spectrometry (MS), measurements and reporting of endogenous androgenic anabolic agents, and analytical requirements for the Athlete Biological Passport (ABP).

Vincent Desfontaine, Jean-Luc Veuthey & Davy Guillarme, School of Pharmaceutical Sciences, University of Geneva, University of Lausanne, Switzerland
Raul Nicoli, Tiia Kuuranne, Swiss Laboratory for Doping Analyses, University Center of Legal Medicine, Lausanne-Geneva, Centre Hospitalier Universitaire Vaudois, University of Lausanne, Switzerland

https://doi.org/10.1515/9783110500776-005

Urine and blood are considered as the matrices of choice for routine doping control analysis. The majority of anti-doping controls is carried out on urine samples, since the collection is noninvasive and large volumes of matrix are available. On the contrary, blood collection is much more invasive with limited sample volume and the transfer of the samples requires more strict protocols regarding time and temperature. Even if the percentage of blood testing is continuously increasing, this is still not the reference matrix for doping control analysis, except for human growth hormone (hGH) assays (serum) and the hematological module of the ABP (whole blood) [3].

The analytical process for the detection of presence or absence of a doping agent in urine is routinely carried out through the schematic workflow described in Figure 5.1. In the laboratory, both A and B samples are registered and managed typically in an appropriate laboratory information management system. The A-bottle is opened and distributed in different aliquots, while the B-bottle is kept frozen with the seal intact. An initial testing procedure (screening) is then carried out on the A-sample aliquots, followed by a confirmation procedure, if required. The screening procedure must be sufficiently fast, selective, and sensitive to avoid false-negatives and to minimize the risk of false-positive results. In the case of a presumptive analytical finding in the screening step, a confirmation procedure focused on the suspected target substance, including some possible metabolite(s) should then be performed, and the final result is reported only after the confirmation procedure. The final analytical step following the adverse analytical finding (positive) A-sample report is the confirmation procedure in the B-sample, unless the athlete waives her/his right to.

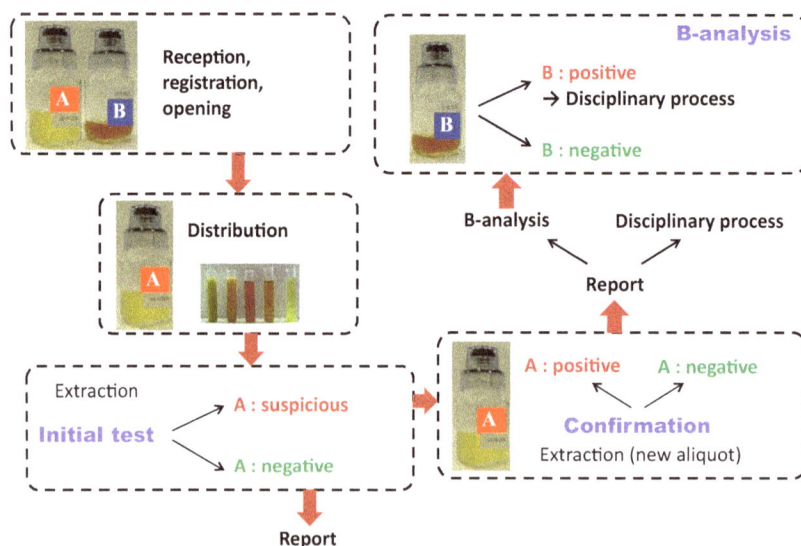

Figure 5.1: Typical workflow of doping control analysis.

5.2 Chemical diversity of doping agents

Doping control analysis includes the identification and, in some few cases, the quantification of target compounds in biological fluids. The prohibited substances and methods are defined by WADA (see Table 5.1) [4], and they need to be tracked in different specimens (urine and blood) collected in- and out-of-competition test events. The prohibited compounds are classified in ten classes (S0–S9), while the illicit methods are classified in three different categories (M1–M3). It is important to notice that alcohol (P1) and β-blockers (P2) are prohibited only in a limited number of sports. A description of the different classes of doping agents is provided in Table 5.1, including some examples of specific substances. Compounds are divided into (i) nonthreshold substances, for which an adverse analytical finding involves their identification in biological matrix, and (ii) threshold substances, such as ephedrine and derivatives, salbutamol, and carboxy-Tetrahydrocannabinol, which are prohibited only above a certain concentration, which requires quantitative determination in biological sample. To ensure and harmonize adequate sensitivity level in all accredited anti-doping laboratories, Minimum Required Performance Levels (MRPLs) for analytical methods have been established by the WADA. The MRPL corresponds to the minimum performance that should be obtained for the detection of nonthreshold substances.

The forbidden substances cover a broad chemical diversity and a wide range of physicochemical properties. The list includes some very polar and zwitterionic substances such as meldonium and 5-Aminoimidazole-4-carboxamide ribonucleotide (AICAR), some less polar but still ionizable compounds such as stimulants, diuretics, and narcotics (e.g., amphetamine and furosemide), and also a large number of highly lipophilic substances such as exogenous and endogenous anabolic steroids (e.g., oxandrolone and testosterone). Some gases (xenon and argon) were also recently added to the WADA list. Finally, there is also a growing number of peptides (e.g., Growth hormone-releasing peptides (GHRPs) and insulin-like growth factor 1 (IGF-1)) and proteins (hGH and erythropoietin) which also need to be detected in biological fluids. In total, there are approximately 250–300 compounds included in the WADA list and this number has continuously increased over the last few years [5, 6]. In order to keep the list dynamic and flexible for new chemical entities, most of the categories are not closed but include also "other substances" with a similar chemical structure or similar biological effect(s).

5.3 Analytical trends in doping control analysis

Due to the important chemical diversity of the prohibited substances, anti-doping laboratories have to apply multiple analytical techniques, including immunological, biochemical, and chromatography-MS methods to the routine analyses. In this

Table 5.1. Summary of the 2017 WADA prohibited list.

Categories	Categories description	Prohibition	Subcategories/specific examples/observations*
S0	Nonapproved substances	In- and out of competition	Any pharmacological substance with no current approval for human therapeutic use (e.g., drugs under preclinical or clinical development or discontinued, designer drugs, substances approved only for veterinary use)
S1.1.a	Exogenous anabolic androgenic steroids		Bolasterone, calusterone, clostebol, dehydrochloromethyltestosterone, drostanolone, fluoxymesterone, mesterolone, methandienone, methasterone, methyltestosterone, oxandrolone, stanozolol, trenbolone, etc.
S1.1.b	Endogenous anabolic androgenic steroids		19-Norandrostenediol, 19-Norandrostenedione, androstenediol, androstenedione, boldenone, boldione, DHT, DHEA, nandrolone, testosterone and their metabolites and isomers
S1.2	Other anabolic agents		Clenbuterol, SARMs (e.g., andarine and ostarine), tibolone, zeranol, zilpaterol
S2	Peptide hormones, growth factors, related substances, and mimetics		Erythropoietin-receptor agonists (e.g., EPO, EPO-Fc, peginesatide, and ARA-290), HIF stabilizers and activators (FG-4592, argon, xenon), hCG and LH, corticotrophins, growth hormone and their releasing factors, other growth factors, etc.
S3	β2-Agonists		Bambuterol, fenoterol, formoterol, higenamine, indacaterol, olodaterol, procaterol, reproterol, salbutamol, salmeterol, terbutaline, vilanterol, etc.
S4	Hormone and metabolic modulators		AICAR, androstatriendione, arimistane, clomiphene, cyclofenil, exemestane, formestane, fulvestrant, GW1516, insulins, letrozole, meldonium, raloxifene, tamoxifene, testolactone, toremifene, trimetazidine, etc.
S5	Diuretics and masking agents		Acetazolamide, amiloride, bendroflumethiazide, canrenone, chlortalidone, desmopressin, furosemide, glycerol, HES, hydrochlorothiazide, indapamide, spironolactone, triamterene, tolvaptan, etc.

(continued)

Table 5.1 (Continued)

Categories	Categories description	Prohibition	Subcategories/specific examples/observations*
S6	Stimulants	In competition	Benzfetamine, cathine, cathinone, mephedrone, methedrone, ephedrine, etilefrine, heptaminol, isometheptene, MDA, MDMA, methylephedrine, methylphenidate, MHA, oxilofrine, pseudoephedrine, sibutramine, tuaminoheptane, etc.
S7	Narcotics		Buprenorphine, dextromoramide, fentanyl and derivatives, heroin, hydromorphone, methadone, morphine, nicomorphine, oxycodone, oxymorphone, pentazocine, pethidine, etc.
S8	Cannabinoïds		Cannabis, hashish, marijuana, Δ9-tetrahydrocannabinol, JWH-018, JWH-073, HU-210, etc.
S9	Glucocorticoïds		Beclomethasone, beta/dexamethasone, budesonide, ciclesonide, deflazacort, desonide, fludrocortisone, flumethasone, flunisolide, methylprednisolone, prednisolone, prednisone, triamcinolone acetonide, etc.
P1	Alcohol	In competition-in particular sports*	Detection conducted by analysis of breath and/or blood. The doping violation threshold is equivalent to a blood alcohol concentration of 0.10 g/L.
P2	β-Blockers**		Acebutolol, alprenolol, atenolol, betaxolol, bisoprolol, bunolol, carteolol, carvedilol, celiprolol, esmolol, labetalol, levobunolol, metipranolol, metoprolol, nadolol, oxprenolol, pindolol, propranolol, sotalol, timolol, etc.
M1	Manipulation of Blood and Blood Components	In- and out of competition	Autologous, homologous, and heterologous blood transfusions, perfluorochemicals, RSR13, HBOCs, etc.
M2	Chemical and physical manipulation		Tampering or attempting to tamper, urine substitution, urine adulteration, intravenous infusions and/or injections of more than 50 mL per 6 h period
M3	Gene doping		The transfer of polymers of nucleic acids or nucleic acid analogues; the use of normal or genetically modified cells

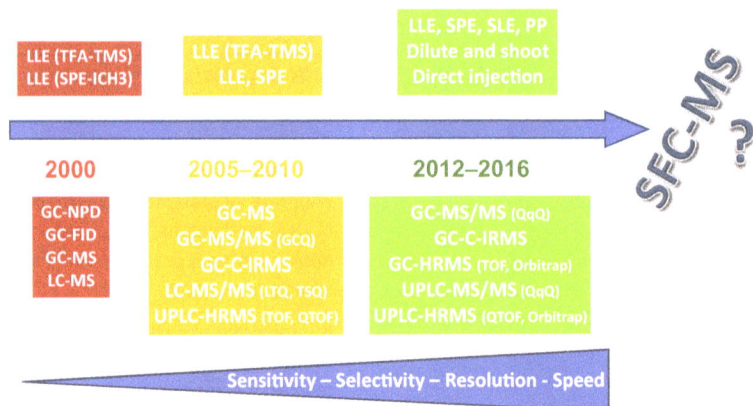

Figure 5.2: Analytical trends in doping control analysis.

chapter, only chromatographic-MS methods will be described. The evolution of analytical instrumentation and methods from 2000 to 2016 (sample preparation, chromatography, and MS) is provided in Figure 5.2 [3]. The developed methods focus not only on the direct detection of parent compounds, but also on the determination of their major phase I and phase II metabolites, to further improve the detection windows capabilities in the matrix of interest [7].

Sample preparation is an important part of the analytical process, due to the complexity of the biological matrices (blood and urine), which may contain important amount of salts, lipids, proteins, and other endogenous compounds. Sample preparation is therefore often mandatory to achieve a sufficient sensitivity and selectivity, and also to limit the contamination of the analytical instruments. Various sample preparations can be used, but, traditionally, liquid–liquid extraction (LLE) has been applied most widely. However, since LLE is tedious and time-consuming, it was progressively replaced by solid-phase extraction which can be automated more easily and also offers some high recoveries for a wide range of analytes. Then, considering the high selectivity and sensitivity of modern MS detectors, sample preparation was further simplified for the initial screening procedure, and less or even nonselective procedures can also be currently employed, especially in LC-MS-based methods. For urine samples, the dilute-and-shoot approach is highly generic, fast, inexpensive, and almost does not require any equipment. However, the sensitivity is reduced due to the dilution factor and also by the matrix-dependent ion suppression [8]. For blood samples, a simple and generic protein precipitation can also be applied using an appropriate solution [9].

In terms of chromatographic techniques, both GC and LC are considered as reference methods. Their effectiveness benefits from the recent excellent technical improvements in terms of selectivity and rapidity. GC-flame ionization detector and GC-MS quickly became the standard instrumentation for the detection and quantification of illicit substances. In GC, hydrolysis and derivatization steps are required prior to the analysis of doping agents, to ensure volatility but also to improve their sensitivity. These additional procedures may induce variability and are considered time-consuming and difficult to automate. In addition, due to the recent improvements of LC (the commercial introduction of systems compatible with pressure up to 1,000–1,500 bar and columns packed with sub-2 µm fully porous particles (ultrahigh-performance liquid chromatography [UHPLC] as well as sub-3 µm superficially porous particles), the performance of LC has drastically improved. Analysis times in the range 3–10 min can easily be achieved and the hydrolysis step, in some cases, can be omitted to directly target intact phase II metabolites with increased sensitivity (e.g., stanozolol glucuronides [10]). Therefore, nowadays UHPLC-MS represents the analytical platform of choice for various classes of doping substances. As described in this chapter, supercritical fluid chromatography (SFC)-MS is still scarcely used for routine doping control analysis, but due to certain obvious advantages, the situation can be expected to evolve in a near future (see later).

Finally, MS has rapidly emerged as the gold standard detection mode in anti-doping analysis, thanks to its high selectivity and sensitivity, and also due to its superior identification capability based on mass spectral information on the target compounds [11]. In the beginning of the 2000s, low-resolution single quadrupole instruments were mostly used, but this technology has rapidly been replaced by triple quadrupole instruments, to improve selectivity and sensitivity. Today, high-resolution MS systems, such as time-of-flight-based instruments (TOF and quadrupole-quadrupole TOF) or Orbitrap, are becoming more and more popular. Owing to the scanning properties and nonrestricted data acquisition of the new instruments, it is possible to review previously acquired data, once a new doping agent (or a new metabolite) is discovered. The versatility and possibility of retrospective analysis is a major reason for the increasing popularity of TOF and Orbitrap instruments for doping control analysis [3].

5.4 Combining SFC with MS

Due to the physicochemical properties and compressibility of the mobile phase employed in SFC, directing the column effluent into the electrospray ionization source (ESI) is more difficult than in LC. Therefore, various SFC-MS interfaces were tested and commercialized [12]. The available interfaces for hyphenating SFC and MS should be first able to manage the compressibility of SFC mobile phase and preserve the chromatographic integrity (i.e., maintaining retention, selectivity, and efficiency). Furthermore, when the SFC mobile phase is not anymore under pressure, the mobile phase density is reduced, and the compounds may precipitate. The interface should also be able to circumvent this potential issue [13]. Finally, when using an SFC mobile phase containing a high proportion of CO_2, the poor ESI ionization yield can be enhanced thanks to the addition of a sheath liquid (mostly composed of pure methanol).

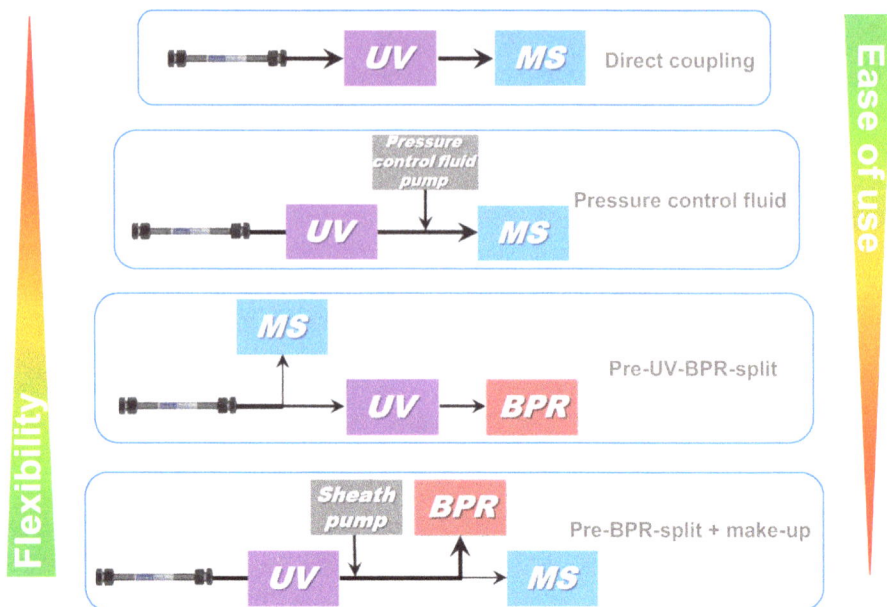

Figure 5.3: Available SFC-MS interfaces.

As illustrated in Figure 5.3, various commercial SFC-MS interfaces now allow robust and highly sensitive SFC-MS operation. Each of these interfaces may have noticeable effects on chromatographic reliability, flexibility, sensitivity, and user-friendliness.

Nowadays, the *pre*-backpressure regulator (*BPR*)-*split* + *make-up* interface is the most widely used on modern SFC-MS instruments and is commercially available from several providers (Agilent, Waters, and Shimadzu) [14]. In this interface, the UV detector is located at the column outlet. Two zero-dead-volume T-unions are located between the UV and MS detectors. The first T-union allows the addition of a make-up liquid (often a protic solvent such as methanol, which could also contain some acidic or basic additives to promote ionization) which is generally delivered by an isocratic pump. This liquid is mixed to the chromatographic effluent to avoid analyte precipitation during CO_2 decompression and enhance the ionization efficiency at low MeOH proportion. The second T-union is used for flow splitting. Indeed, part of the flow is directed to the MS, while the rest of the flow is directed to the BPR. Because the BPR is placed downstream of the SFC column, the chromatographic integrity (retention, selectivity, efficiency, peak shape, and resolution) is completely maintained. In some cases, this interface is also heated at around 60 °C, to limit as much as possible the cooling effect of the eluent expansion at the column outlet. This decreases peak broadening which is sometimes observed in SFC-MS, because of the decompression cooling [14]. This *pre-BPR-split + make-up* interface was reported as the most reliable in terms of retention times and peak areas reproducibility. Due to its high sensitivity, linearity, and robustness, this configuration is therefore recommended for both qualitative and quantitative analyses [15].

5.5 SFC-MS for the screening of relatively polar substances

Until now, SFC-MS has been scarcely used for doping control purposes, despite some clear advantages of this chromatographic technique [16]: (i) the low mobile phase viscosity and high diffusion coefficients in SFC provide high kinetic performance; (ii) the organic solvent consumption remains reasonable, and in some cases a proportion of organic modifier up to 40% can be used in modern SFC; (iii) a wide range of substances, from polar to highly nonpolar, can be analyzed in SFC, with the same mobile phase components (CO_2 and methanol). SFC has made a remarkable comeback in the past few years due to significant improvements in instrumentation (high performance, reliability and robustness) and column technology [17, 18], allowing its use in quality-controlled routine environment such as anti-doping laboratories. In the last few years, a few papers were released, describing the use of SFC for doping control analysis. Parr et al. [19] have recently reported the use of SFC-MS/MS for the screening of about 35 polar sympathomimetic analytes and their metabolites which were poorly retained in reversed phase liquid chromatography (RPLC) conditions. In this study, etilefrine and octopamine, showing almost no retention under RPLC conditions, were sufficiently retained in SFC. The developed SFC method also demonstrated applicability to sulfoconjugates analysis, which generally display very low retention under RPLC and Hydrophilic interaction chromatography (HILIC) conditions. This enhanced retention was attributed to the polar nature of the stationary phase employed in SFC (2-ethylpyridine phase), in combination with MeOH as mobile phase component and ammonium acetate as additive.

The significant changes in retention observed in SFC were also highlighted in another study, describing the behavior of 110 doping agents (acidic, basic, and neutral substances) belonging to various classes from the WADA list (i.e., narcotics, stimulants, diuretics, and β-blockers) [20, 21]. The experiments were performed in RPLC-MS/MS (reference method using a Waters Acquity BEH C18 50 × 2.1 mm, 1.7 μm column) and SFC-MS/MS (new method using a Waters Acquity UPC² BEH 100 × 3 mm, 1.7 μm column), using the same MS/MS instrument. As shown in Figure 5.4, very diverse retentions and selectivities were obtained in RPLC and SFC, demonstrating the complementarity of these two analytical strategies. As shown in this figure, the poorly retained compounds in RPLC (highlighted in yellow in the right figure) were sufficiently retained in SFC, and the reverse situation (highlighted in red in the left figure) was also true. This was attributed to the fact that the retention mechanisms were quite different between both modes (retention mostly occurs through hydrophobic interactions and hydrogen bonding in RPLC and SFC, respectively). Finally, it was possible to use SFC for analyzing a much wider range of compounds (including the most polar doping agents) compared to RPLC. Moreover, the elution window was wider in SFC, allowing a better separation of the analytes.

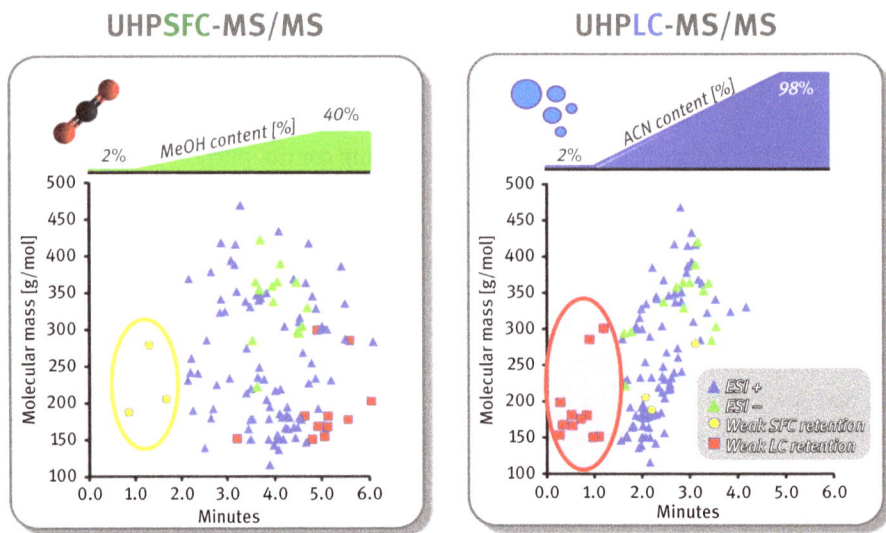

Figure 5.4: Elution distribution of 114 doping agents under UHPLC–MS/MS and UHPSFC–MS/MS conditions. The UHPLC separation was performed on a Waters Acquity BEH C18 50 × 2.1 mm, 1.7 μm column, while the UHPSFC analysis was conducted on a Waters Acquity UPC² BEH 100 × 3 mm, 1.7 μm column. The compounds ionized in ESI– are displayed in green (triangles), while those ionized in ESI+ are displayed in blue (triangles) under UHPLC–MS/MS and UHPSFC–MS/MS conditions, respectively. Reprinted from Ref. [21]. Copyright 2014, with permission from Elsevier.

Besides retention, Figure 5.4 also shows that selectivity was strongly different between the two chromatographic modes, demonstrating again the complementarity of these two chromatographic modes. It is also important to notice that the 110 selected doping agents were successfully analyzed in both RPLC and SFC, using generic mobile phase conditions (from 2 to 40% MeOH in SFC and from 2 to 98% ACN in RPLC). Acceptable peak shapes and MS detection capabilities were obtained within an analysis time of only 7 min, enabling the application of these two methods for screening purposes.

In terms of achievable sensitivity, the ESI source settings were similar in SFC-MS/MS and LC-MS/MS. The presence of volatile salts in the mobile phase (i.e., 10 mM ammonium formate) did not reduce sensitivity in SFC-MS/MS, as demonstrated earlier [20], but it could improve the peak shapes of ionizable compounds. In terms of make-up solvent, pure methanol provided the highest sensitivity. The Limits of detection (LODs) were compared in LC and SFC-MS/MS for the 110 doping agents in urine matrix, after a simple *dilute and shoot* pretreatment [21]. When using an old-generation triple quadrupole device, the sensitivity was similar between LC and SFC for 27% of the doping agents, while it was improved in SFC-MS for 65% of the compounds. The better sensitivity achieved in SFC-MS was probably related to a better mobile phase desolvation due to the absence of water in large proportion in SFC-MS. However, when using a more recent MS/MS instrument of the same brand, the sensitivity improvement was much lower, with 46% of doping agents offering similar sensitivity in LC-MS/MS and SFC-MS/MS, and only 38% of the doping agents with improved LODs in SFC versus LC. This behavior could be attributed to a better compatibility of modern MS/MS with highly aqueous RPLC mobile phases [21]. In SFC-MS/MS, it is finally hard to assess whether the sensitivity could be improved compared to LC-MS/MS, only based on the structure of the compounds, since there are numerous additional parameters that also contribute to the change in MS signal. However, in many cases, the SFC-MS/MS sensitivity was at least equivalent or even better than the one achieved in LC-MS/MS for the selected group of doping agents.

Another important parameter to consider when analyzing biological fluids such as urine is the relevance of matrix effects, as it can lead to poor accuracy, precision, and sensitivity of the method. To date, the number of studies investigating matrix effects in SFC-MS is relatively limited. In the case of doping control analysis, the occurrence of matrix effects was recently evaluated in SFC-MS versus LC-MS for urine matrices [21]. The incidence of matrix effects was found to be somewhat lower in SFC, and this observation was independent from the concentration. In average, about 50% of the doping agents were not affected by matrix effects in LC-MS, while this number was equal to 70% in SFC-MS. Obviously, the salts and polar endogenous compounds contained in urine were certainly responsible for matrix effects. These critical compounds were poorly retained in LC-MS, while they eluted much later in SFC-MS, due to the very different retention mechanisms. This explains the difference in matrix effects between the two chromatographic techniques.

Finally, it is also worth mentioning that an SFC-MS/MS method was also recently proposed for the enantiomeric separation of (*R*)- and (*S*)-clenbuterol (a bronchodilator sympathomimetic doping agent) to distinguish between abuse and exposure to contaminated meat [22].

5.6 SFC-MS for the screening of anabolic androgenic steroids and related substances

Among the WADA list of prohibited substances, the determination of the numerous anabolic agents (particularly androgenic steroids) is particularly demanding, as MRPLs are extremely low. In addition, there is a significant number of isomers and metabolites (steroids share very similar backbone structures, with minor differences related to the ring saturation and the presence and position of additional functional groups), which are difficult to discriminate [23]. Based on these critical features, GC-MS(MS) remains the gold standard for analyzing these substances in anti-doping laboratories. However, to make the substances sufficiently volatile and improve their detection limits, LLE and derivatization steps are required prior to the analysis [24, 25].

Remarkably, SFC is known as a reference technique for the analytical identification and characterization of various types of steroids [26, 27], but has only been scarcely applied for the determination of steroids and their derivatives in anti-doping analysis [28–30], and still not yet in routine analyses.

In a recent study [29], the authors have developed an analytical method for the high-throughput screening of approximately 100 substances (parent compounds and phase I metabolites) in urine. All these substances belong to the complex classes of anabolic androgenic steroids, synthetic cannabinoids, hormones and metabolic modulators, and glucocorticoids which include a lot of isobaric compounds that have to be detected at low concentrations in urine. In the developed method, the extraction of compounds of interest from urine was carried out, after enzymatic hydrolysis, *via* a high-throughput supported-liquid extraction (SLE) method in the 48-well plate format. Then, the chromatographic separation of the 100 substances was performed in SFC, after a careful optimization of the chromatographic conditions to achieve acceptable peak shapes and selectivity, taking into account the presence of a high number of interferences related to steroid isomers and the ions produced by the loss of water in the electrospray source. Among the optimization tasks, the choice of SFC separation conditions was the most challenging. As shown in Figure 5.5, some obvious differences were observed between column chemistries

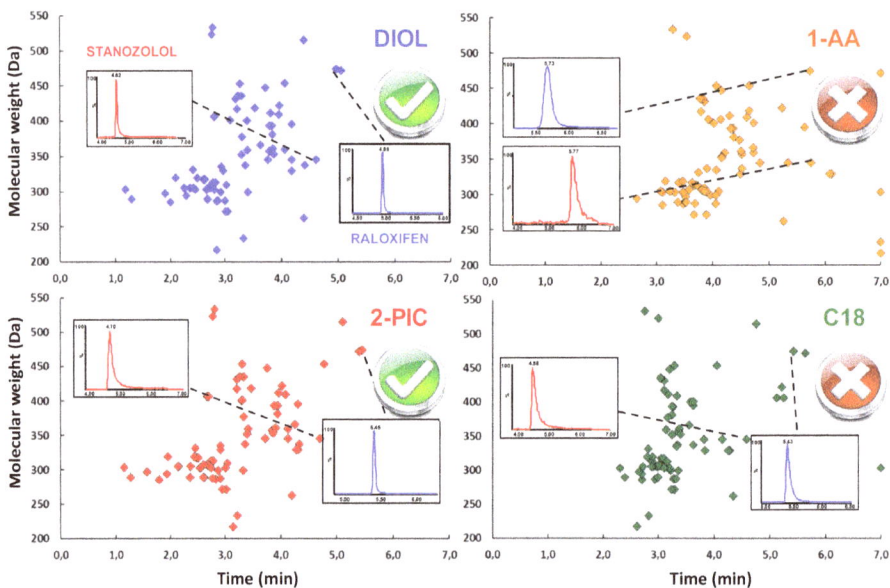

Figure 5.5: Selectivity charts for four columns. Each doping agent is plotted according to its retention time (X-axis) and mass (Y-axis) for the diol, 2-PIC, and 1-AA, and HSS C18 SB columns. The chromatograms of two compounds (3'OH-stanozolol in red and raloxifen in blue) are shown to highlight peak shape performance on each column. Reprinted from Ref. [29]. Copyright 2015, with permission from Elsevier.

in SFC, and the elution of all the substances was possible with narrow peaks and limited interference issues (high chromatographic selectivity) in only 8 min, using the diol stationary phase chemistry. Finally, the detection was performed with a triple quadrupole MS/MS instrument. The LODs achieved with this method were compared to MRPLs fixed by WADA, for all these classes of compounds. Thanks to the preconcentration factor offered by SLE (fivefold), and the good sensitivity provided by modern SFC-MS/MS platform, LODs were always lower than the MRPLs, except for one metabolite of each of these substances: norethisterone, norbolethone, and turinabol. These results prove that the SLE-SFC-MS/MS method is able to offer excellent performance in terms of turnaround, selectivity, and sensitivity. It could potentially be employed in anti-doping laboratories for screening purposes of exogenous anabolic androgenic steroids, synthetic cannabinoids, hormones and metabolic modulators, and glucocorticoids in a near future, as a replacement solution of the reference strategy, namely LLE-GC-MS/MS.

In another study [30] from the same authors, the performance of the SLE-SFC-MS/MS method was compared to an SLE-LC-MS/MS method and to the reference LLE-GC-MS/MS routine procedure for the screening of 43 anabolic agents in human urine. Some urine samples spiked with the 43 compounds of interest at different concentration levels were analyzed using the three analytical platforms. As illustrated in Figure 5.6, the complementarity of the techniques was very good and coelutions

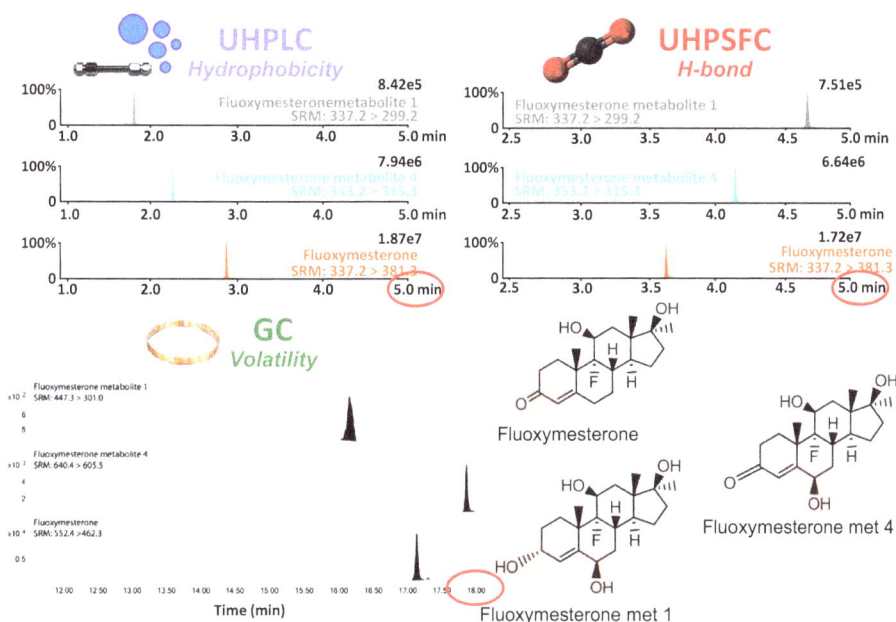

Figure 5.6: Chromatograms for injection of urine spiked with fluoxymesterone and two metabolites at 10 ng/mL in LC–MS/MS, SFC–MS/MS, and GC-MS/MS. Reprinted from Ref. [30]. Copyright 2016, with permission from Elsevier.

observed in LC-MS/MS were solved in SFC-MS/MS and vice versa. As an example, the retention of three compounds (fluoxymesterone and two phase I metabolites having one or two hydroxyl groups) was observed to better understand the chromatographic behavior. As expected, in RPLC the three substances were eluted based on their hydrophobicity (metabolite with two hydroxyl groups < metabolite with one hydroxyl group < parent molecule). In SFC, the elution order was completely reversed since the molecules were eluted taking into account the number of H-bond donor groups on the molecule. Finally, the elution order was again different in GC since the molecules were eluted based on their volatility (metabolite with two hydroxyl groups < parent molecule < metabolite with one hydroxyl group). However, it is also clear that SFC and LC offer some advantages over GC, including a twofold increase in analysis turnover rate compared to the current routine GC analysis (8 min in LC and SFC *versus* 20 min in GC) and above all much faster sample preparation. Due to the good complementarity of the three analytical methods, two of them could be perfectly suitable as screening and confirmation steps of anti-doping analysis procedure.

The limits of detection obtained in SFC-MS/MS were compared to the ones achieved in LC-MS/MS and GC-MS/MS for the 43 anabolic agents in urine [30]. As shown in Figure 5.7, the LC-MS/MS method offers the highest sensitivity for this class of compounds, with 98% of the doping agents detected at the lowest concentration level of the study (0.1 ng/mL). The sensitivity of SFC-MS/MS was also very good, with three quarters of the doping agents also detected at the lowest tested concentration (0.1 ng/mL). In GC-MS/MS, only 14% of the substances were detected at this concentration level. In average, the sensitivity provided by LC and SFC-MS/MS was about 10 times higher than the GC-MS/MS. Therefore, SFC-MS/MS can be considered as a potential routine method for the screening of exogenous anabolic agents in urine in replacement of the reference GC-MS/MS method.

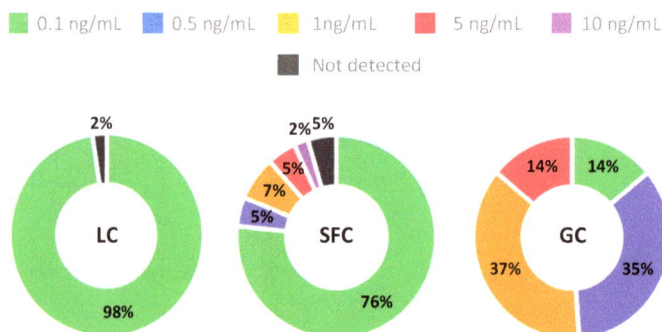

Figure 5.7: Achievable sensitivities in SFC, LC, and GC. Percentage of the 43 anabolic agents detected at 0.1 ng/mL (green), 0.5 ng/mL (blue), 1 ng/mL (orange), 5 ng/mL (red), 10 ng/mL (purple), or not detected (gray) in extracted urine in LC-MS/MS, SFC-MS/MS, and GC-MS/MS. Reprinted from Ref. [30]. Copyright 2016, with permission from Elsevier.

5.7 Conclusion

In conclusion, SFC-MS appears as a promising strategy for the determination of numerous doping agents. Until now, more than 200 compounds from the WADA list were successfully analyzed in SFC. This chromatographic strategy can therefore be considered as a faster, greener, more universal, and orthogonal analytical method, compared to LC-MS and GC-MS. However, the number of studies showing a systematic evaluation of sensitivity, matrix effects, linearity, precision, and accuracy with different biological fluids in SFC-MS for a wide range of doping agents is still too limited. This technique will certainly become, in a near future, a routine analytical strategy in doping control laboratories after having proved its feasibility.

Besides small molecules included in the WADA list, there are also numerous peptides and proteins which also have to be determined in biological fluids. For such large molecules, however, SFC is much less potent than LC, despite a few successful attempts in SFC conditions in the last few years [31, 32].

Finally, SFC is widely used in the pharmaceutical industry at the preparative scale (prepSFC) to purify desired quantities of target compounds. Supercritical carbon dioxide has also been largely used as a sample preparation technique for the extraction of substances from complex matrixes (SFE). Although these two techniques (prepSFC and SFE) have not yet been employed in doping control analysis, they could be an asset to the field in a near future.

References

[1] World Anti-Doping Code, World Anti-Doping Agency, Montreal, 2015.
[2] International Standards for Laboratories, version 8.0, World Anti-Doping Agency, Montreal, 2015.
[3] Nicoli R, Guillarme D, Leuenberger N, Baume N, Robinson N, Saugy M, Veuthey JL Analytical strategies for doping control purposes: Needs, challenges and perspectives. Anal Chem 2016, 88, 508–523.
[4] The 2017 Prohibited List, World Anti-Doping Agency, Montreal, 2017.
[5] Thevis M, Kuuranne T, Geyer H, Schanzer W. Annual banned-substance review: Analytical approaches in human sports drug testing. Drug Test Anal 2014, 6, 164–184.
[6] Thevis M, Kuuranne T, Geyer H, Schanzer W. Annual banned-substance review: Analytical approaches in human sports drug testing. Drug Test Anal 2015, 7, 1–20.
[7] Badoud F, Guillarme D, Boccard J, Grata E, Saugy M, Rudaz S, Veuthey JL. Analytical aspects in doping control: Challenges and perspectives. Forensic Sci Int 2011, 213, 49–61.
[8] Deventer K, Pozo OJ, Verstraete AG, Van Eenoo P. Dilute-and-shoot-liquid chromatography-mass spectrometry for urine analysis in doping control and analytical toxicology. Trends Anal Chem 2015, 55, 1–13.
[9] Bruce SJ, Tavazzi I, Parisod V, Rezzi S, Kochhar S, Guy PA. Investigation of human blood plasma sample preparation for performing metabolomics using ultrahigh performance liquid chromatography/mass spectrometry. Anal Chem 2009, 81(9), 3285–3296.

[10] Schänzer W, Guddat S, Thomas A, Opfermann G, Geyer H, Thevis M, Expanding analytical possibilities concerning the detection of stanozolol misuse by means of high resolution/ high accuracy mass spectrometric detection of stanozolol glucuronides in human sports drug testing. Drug Test. Anal 2013, 5(11–12), 810–818.

[11] Musenga A, Cowan DA, Use of ultra-high pressure liquid chromatography coupled to high resolution mass spectrometry for fast screening in high throughput doping control. J Chromatogr A 2013, 1288, 82–95.

[12] Desfontaine V, Veuthey JL, Guillarme D, Hyphenated detectors: Mass spectrometry [In] Supercritical fluid chromatography, C.F. Poole, Editor, Elsevier, 2017, 213–244

[13] Kott L. An overview of supercritical fluid chromatography mass spectrometry (SFC-MS) in the pharmaceutical industry. Am Pharm Rev 2013, 16, 26–29.

[14] Grand-Guillaume Perrenoud A, Veuthey JL, Guillarme D. Coupling state-of-the-art supercritical fluid chromatography and mass spectrometry: From hyphenation interface optimization to high-sensitivity analysis of pharmaceutical compounds. J Chromatogr A 2014, 1339, 174–184.

[15] Dunkle M, Vanhoenacker G, David F, Sandra P. Agilent 1260 Infinity SFC/MS Solution – Superior sensitivity by seamlessly interfacing to the Agilent 6100 Series LC/MS system. Agilent application note 2011:5990–7972EN.

[16] Lesellier E, West C. The many faces of packed column supercritical fluid chromatography – A critical review. J. Chromatogr A 2015, 1382, 2–46.

[17] Desfontaine V, Novakova L, Guillarme D. SFC–MS versus RPLC–MS for drug analysis in biological samples. Bioanalysis 2015, 7, 1193–1195.

[18] Desfontaine V, Guillarme D, Francotte E, Novakova L. Supercritical fluid chromatography in pharmaceutical analysis. J Pharm Biomed Anal 2015, 113, 56–71.

[19] Parr MK, Wuest B, Naegele E, Joseph JF, Wenzel M, Schmidt AH, Stanic M, De la Torre X, Botre F. SFC-MS/MS as an orthogonal technique for improved screening of polar analytes in anti-doping control. Anal Bioanal Chem 2016, 408, 6789–6797.

[20] Novakova L, Grand-Guillaume Perrenoud A, Nicoli R, Saugy M, Veuthey JL, Guillarme D. Ultra high performance supercritical fluid chromatography coupled with tandem mass spectrometry for screening of doping agents. I: Investigation of mobile phase and MS conditions. Anal Chim Acta 2015, 853, 637–646.

[21] Novakova L, Rentsch M, Grand-Guillaume Perrenoud A, Nicoli R, Saugy M, Veuthey JL, Guillarme D. Ultra high performance supercritical fluid chromatography coupled with tandem mass spectrometry for screening of doping agents. II: Analysis of biological samples. Anal Chim Acta 2015, 853, 647–659.

[22] https://wada-main-prod.s3.amazonaws.com/resources/files/review_s_sterk_11a18ss_0.pdf, consulted in June 2017.

[23] Schanzer W, Thevis M, Human sports drug testing by mass spectrometry. Mass Spec Rev 2017, 36, 16–46.

[24] Massé R, Ayotte C, Dugal R. Studies on anabolic steroids: I. Integrated methodological approach to the gas chromatographic-mass spectrometric analysis of anabolic steroid metabolites in urine. J Chromatogr B 1989, 489, 23–50.

[25] Marcos J, Pascual JA, De la Torre X, Segura J. Fast screening of anabolic steroids and other banned doping substances in human urine by gas chromatography/tandem mass spectrometry. J Mass Spectrom 2002, 37, 1059–1073.

[26] Xu X, Roman JM, Veenstra TD, Van Anda J, Ziegler RG, Issaq HJ. Analysis of fifteen estrogen metabolites using packed column supercritical fluid chromatography-mass spectrometry. Anal Chem 2006, 78, 1553–1558.

[27] Nováková L, Chocholouš P, Solich P. Ultra-fast separation of estrogen steroids using subcritical fluid chromatography on sub-2-micron particles. Talanta 2014, 121, 178–186.

[28] Doue M, Dervilly-Pinel G, Pouponneau K, Monteau F, Le Bizec B, Analysis of glucuronide and sulfate steroids in urine by ultra-high-performance supercritical-fluid chromatography hyphenated tandem mass spectrometry. Anal Bioanal Chem 2015, 407, 4473–84.

[29] Novakova L, Desfontaine V, Ponzetto F, Nicoli R, Saugy M, Veuthey JL, Guillarme D, Fast and sensitive supercritical fluid chromatography – tandem mass spectrometry multi-class screening method for the determination of doping agents in urine. Anal Chim Acta 2016, 915, 102–110.

[30] Desfontaine V, Novakova L, Ponzetto F, Nicoli R, Saugy M, Veuthey JL, Guillarme D, Liquid chromatography and supercritical fluid chromatography as alternative techniques to gas chromatography for the rapid screening of anabolic agents in urine. J Chromatogr A, 2016, 1451, 145–155.

[31] Zheng J, Pinkston JD, Zoutendam PH, Taylor LT. Feasibility of supercritical fluid chromatography/ mass spectrometry of polypeptides with up to 40-mers. Anal Chem. 2006, 78, 1535–1545.

[32] Patel MA, Riley F, Ashraf-Khorssani M, Taylor LT. Supercritical fluid chromatographic resolution of water soluble isomeric carboxyl/amine terminated peptides facilitated via mobile phase water and ion pair formation. J Chromatogr A 2012, 1233, 85–90.

Alexander Marziale

6 SFC as a novel approach to assess polarity and identify intramolecular hydrogen bonding

Abstract: In recent years, supercritical fluid chromatography (SFC) has gained tremendous importance in the analysis and isolation of new chemical entities in drug discovery. Thus, SFC has emerged as a key enabling technology for downstream processing after chemical synthesis across the pharmaceutical industry and other industries. Nonetheless, the application of SFC in this context is by far not limited to the isolation of new chemical matter. In this chapter, we are going to introduce and discuss a novel methodology to assess the polarity and cell membrane permeability of low molecular weight drugs and other chemical matter by SFC. The so-called EPSA methodology offers a straightforward tool for profiling drug compounds with respect to permeability and polarity at high throughput.

Key words: supercritical fluid chromatography, EPSA, polarity, cell membrane permeability, intramolecular hydrogen bonding, hydrogen bond(s), polar surface area

6.1 Introduction

Increasing comprehension of biological systems is providing pharmaceutical research in both academia and industry with exciting novel target classes, including protein–protein and protein–nucleic acid interactions, which are often undruggable by classic small molecule approaches. Consequently, a new generation of molecules is necessary to tackle these targets, which led researchers around the globe to push the frontline of synthetic chemistry toward the so-called New Modalities and explore chemical space beyond classic small molecules and Lipinski's rule of five (RO5) [1]. These new modalities comprise nucleic-acid-based therapeutics such as antisense oligonucleotides and ribozymes, albumin binders, proteolysis targeting chimeric agents, natural-product-derived molecules, structures originating from diversity-oriented synthesis, and novel peptidic agents [1, 2]. While these novel therapeutics allow targeting a far broader biological space they're typically not compliant with the RO5 regime and consequently at odds with good oral absorption [3]. For compounds within this space it is uncertain how physicochemical properties affect cell permeability and designing pharmacologically active compounds with favorable pharmacokinetic profiles and oral bioavailability in vivo represents a major challenge for medicinal chemists [4].

Alexander Marziale, PhD, Novartis Pharma AG Novartis Institutes for Biomedical Research Basel, Switzerland

https://doi.org/10.1515/9783110500776-006

Peptidic drugs, for example, typically possess properties far beyond RO5 space and are delivered by injection due to lack of oral availability [5]. Poor oral adsorption in peptides is associated with proteolytic degradation, high clearance, and reduced permeability due to the overall high polarity [6]. However, the immunosuppressant cyclosporine A exhibits an oral bioavailability of about 30% despite a molecular weight of over 1,200 g/mol, 12 H-bond acceptors, and a polar surface area of 278.8 Å^2 [7]. Extensive studies have revealed that cyclosporine exhibits significant structural flexibility and is capable of adopting a plethora of different confirmations, allowing to form intramolecular hydrogen bonds (IMHBs) and shielding its exposed polarity [8]. This is further corroborated by studies of cyclic tri- and hexapeptides, suggesting that intramolecular hydrogen bonding may contribute to improved membrane permeability [9]. Strategies to improve cell permeability of peptides comprise measures such as N-methylation and masking of H-bond donors through incorporation of unnatural amino acids [10].

Cell permeability is typically monitored through cellular monolayer assays such as Caco-2 and low-efflux Madin Darby canine kidney cells (MDCK-LE), as well as artificial lipid bilayer assays such as PAMPA (parallel artificial membrane permeation assay) [11]. While these assays work well for small molecules, they have significant shortcomings with respect to compounds beyond the RO5 space such as peptidic agents and other new modalities [12].

One property that greatly affects cell membrane permeability is polarity. It is a fundamental physicochemical property and impacts chemical and biological attributes of compounds as well as their three-dimensional shape and conformation. Simultaneously, the exposed polarity of a molecule and its shape are directly influenced by the presence or absence of IMHBs. These noncovalent interactions are ubiquitous among organic molecules and biomacromolecules, and they are critical for the biological activity of these compounds since they define the secondary structure of peptides and proteins, as well as the conformation of small molecules [13]. The importance of IMHBs with respect to membrane permeability and adsorption was discussed previously. Yet, despite the tremendous relevance of IMHBs to drug development projects they are often underrecognized and rarely predicted.

The identification of IMHBs has been investigated by a variety of different analytical and computational approaches in the past. A number of groups have used nuclear magnetic resonance (NMR) to identify IMHBs, either by comparison of chemical shifts in varying solvents, by deuteration experiments, or by measuring temperature coefficients [14]. Also Fourier transform-infrared spectroscopy (FT-IR) was previously used to determine the presence of IMHBs, Winningham and Sogah [15] used a combined approach, employing both FT-IR and variable temperature NMR to distinguish between intramolecularly hydrogen-bonded and free NH groups. Furthermore, a methodology was reported that relies on the comparison of $\Delta\log P$ values of pairs of molecules. This pairwise analysis of $\Delta\log P$ values describes the tendency of compounds to form IMHBs and serves as a molecular descriptor for the prediction of IMHBs in molecules. The methodology is however hampered by the necessity for matched pairs and the low throughput [16].

6.2 Chromatographic methods for IMHB identification

6.2.1 Immobilized artificial membrane high-performance liquid chromatography

In liquid chromatography (LC) the retention of an analyte toward the solid phase is predominantly driven by polarity, and this behavior enables identification of IMHBs between matched pairs of molecules through retention time analysis. In 1984 Clark and coworkers observed enhanced chromatographic retention in *ortho*-substituted benzamides capable of forming an IMHB. For those substituents incapable of internal hydrogen-bond formation reduced retention was found [17]. Today, this chromatographic method is well established in drug development labs and chromatographic columns made of phosphatidylcholine, a major constituent of cell membranes, are used. These immobilized artificial membranes (IAM) mimic the surface of a biological cell membrane and make IAM HPLC (high-performance liquid chromatography) an important tool for high-throughput prediction of drug membrane permeability. The results obtained from this method typically correlate well to traditional in vitro assays such as LE-MDCK and Caco-2 [18]. Nevertheless, IAM HPLC is a reversed-phase chromatographic method and as such utilizes a mobile phase that is composed of water and a water-miscible organic solvent. Consequently, the measurements are performed in a biased environment since water has the ability to disrupt any IMHBs within the compound of interest.

6.2.2 Supercritical fluid chromatography-based IMHB detection and polarity assessment

In an effort to address the shortcomings of methods that are currently used for the detection of IMHB, Goetz and coworkers [19] recently reported the development of a supercritical fluid chromatography (SFC)-based methodology. In SFC the mobile phase consists of supercritical carbon dioxide ($scCO_2$) and a modifier, typically methanol or other alcohols. Technically, SFC is a normal-phase method and the partition of compounds is achieved between a polar stationary phase and a nonpolar eluent. Thus, SFC offers an unbiased environment under nonaqueous conditions with low dielectric constant solvents [20]. Another critical advantage of SFC over RP-HPLC is the increased speed of chromatographic methods, due to the much higher diffusivity and the lower density of $scCO_2$ compared to liquids higher flow rates at lower backpressures can be achieved [21]. This renders SFC an ideal method for high-throughput IMHB detection and polarity assessment. The underlying principle for this approach is once again the assumption that the presence of IMHBs in a molecule reduces its overall polarity compared to a matched pair which is unable to form any IMHBs. This enables a pair-wise analysis since in SFC the retention times of closely related molecules can differ significantly. The authors verified this hypothesis by investigating matched pairs of drug-like compounds with respect to their retention times, and this SFC-based method was named "EPSA" [19].

The authors developed a chromatographic SFC method that allowed for separation of closely related compounds based on their polarity and avoided distortion of IMHBs. A stationary phase, namely the Phenomenex Chirex 3014, a silica-bonded chiral column featuring an (S)-valine moiety bound to (R)-1-(α-naphthyl)-ethylamine through a urea linker (Scheme 6.1) was used.

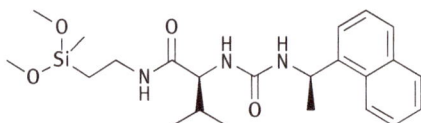

Scheme 6.1: Phenomenex Chirex 3014 stationary phase.

The measurements were performed on an Agilent-Aurora SFC system equipped with a single quadrupole LC-MS (liquid chromatography-mass spectroscopy) detector with ESI source. A 4.6 mm × 250 mm Chirex 3014 column (Phenomenex, Torrance, CA) with 5 μm particles and 100 Å pore size was used at a flow rate of 5 mL/min and a temperature of 40 °C. A linear gradient from 5% to 50% methanol at 5%/min was applied with a total runtime of 13 min. Ammonium formate (20 mM in HPLC-grade methanol) was used as a modifier for the mobile phase. A volume of 5 μL of 3 mM dimethylsulfoxide solution was injected per run.

The authors decided to use the TPSA (topological surface area) value, which is defined as the sum of surface contributions of polar atoms, as a measure for polarity. A correlation between the TPSA and the SFC retention time was established using a set of 118 compounds with a TPSA range of 40–130 Å^2 that were incapable of forming IMHBs. An excellent linear correlation ($R^2 = 0.94$) between TPSA and retention time was found and the authors concluded that SFC could be used as an indirect experimental technique to assess the polarity of compounds [19]. The linear equation that was derived from the plotting of TPSA values against SFC retention times was used to calculate the so-called EPSA [19] values. To test the methodology the authors identified a test set of matched molecular pairs, where one partner has the conformational ability to form IMHBs while the other has not, by mining the Pfizer corporate database. A fully automated structure search was conducted to identify suitable topologies that were defined previously based on their propensity to form IMHBs and their relevance in drug discovery [22]. Further filtering for drug-like properties (clog $P \le 4$, $55 \le$ TPSA ≤ 110, etc.) revealed a total of 84 matched pairs that were used in this study. A few selected topologies are shown in Figure 6.1.

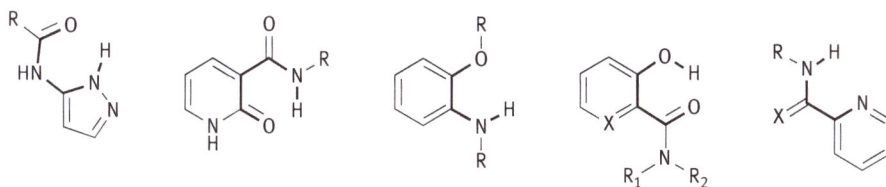

Figure 6.1: Topologies that allow for the formation of IMHBs [19].

A pair-wise analysis was conducted by comparing the difference in EPSA (ΔEPSA) for a matched pair based on the hypothesis that the presence of an IMHB reduces the exposed polarity of a compound relative to its control. Repeated measurements of EPSA values revealed that the obtained values were consistently within ±1 units, thus setting an error bar of ±2 units. Consequently, a ΔEPSA of +3 units was considered significant and a positive indication for the presence of an IMHB [19].

These findings were corroborated by NMR, since this analytical method is very sensitive to structural changes and chemical shifts can be recorded with high accuracy. Proton spectra for all matched pairs were acquired and the chemical shifts of each hydrogen bond donor were recorded whenever feasible. Non C–H proton signals were identified by two-dimensional HSQC experiments. In cases where the hydrogen bond donor proton signal could be assigned ¹H spectra were recorded at various temperatures, allowing for the determination of temperature coefficients. Generally, stronger IMHBs decrease shielding and result in a downfield shift toward higher ppm values [23]. The presence of IMHBs also has an influence on the temperature dependence of hydrogen bond donor proton signals. Consequently, the temperature coefficient can serve as a measure for the presence of intramolecular hydrogen bonding [14c, 24].

Significant ΔEPSA values, as high as 41, were found for a variety of topologies and the corresponding control compounds. Notably, N–H to C=O IMHBs are the most prominent structural motifs among the different IMHB interactions that were evaluated in this study. A number of relevant matched molecular pairs and the respective EPSA values are shown in Figure 6.2.

The ΔEPSA value of 11 units that was found for the matched pair in Figure 6.2a) highlights the existence of an IMHB and is further supported by the downfield shift to 10.07 ppm of the N–H amide proton signal. In this particular example, the *ortho*-substitution pattern allows the formation of an IMHB between the carbonyl and the amide N–H to form a six-membered ring. A different topology is represented by the matched pair in Figure 6.2(b). The EPSA values were found to be 52 and 65 units, while the chemical shift of the N–H amine signals is 9.0 and 5.5 ppm, respectively. Both findings clearly point toward the existence of an IMHB in the *ortho*-substituted aniline derivative. Finally, in Figure 6.2(c) two controls that exhibit *meta-* and *para*-substitution patterns with respect to the pyridine nitrogen atom have been investigated in comparison to the *ortho*-substituted compound. With EPSA values of 106 and 94

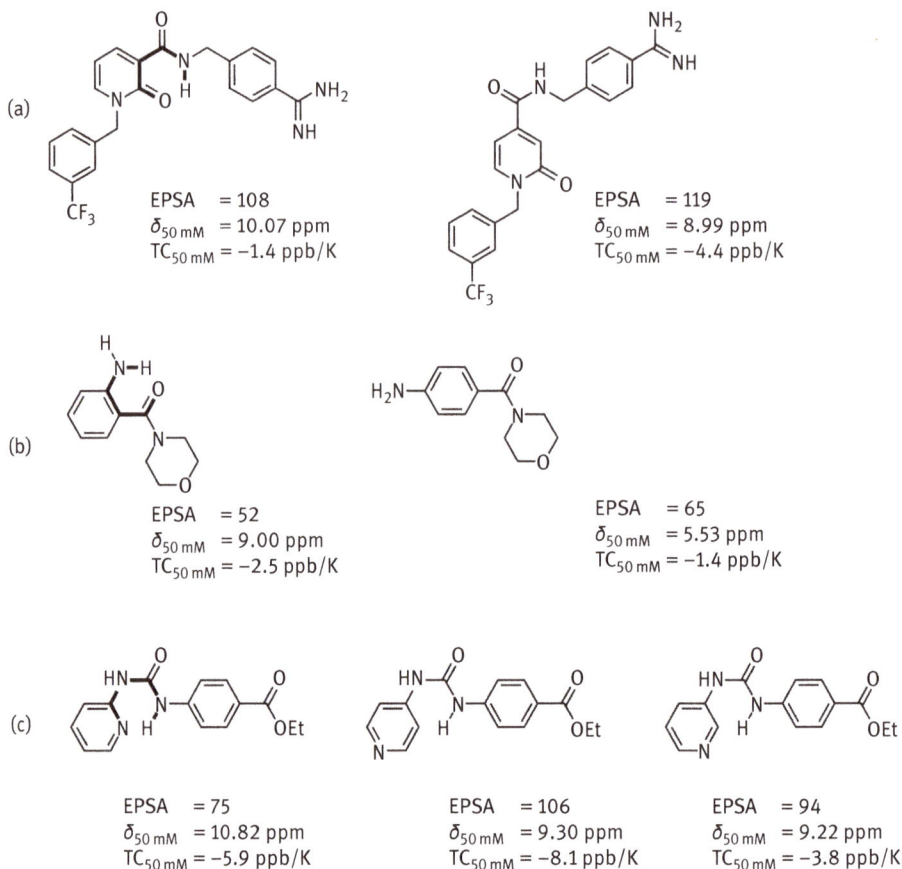

Figure 6.2: Examples of matched pairs of compounds where IMHBs are highlighted by EPSA [19].

both control compounds were found to be considerably more polar than compound X and the chemical shifts for the N–H protons are in line with these observations. However, the temperature coefficient for compound X of –5.9 ppb/K is not consistent with the presence of an IMHB.

However, in a number of cases a ΔEPSA of close to zero was found between matched pairs when the presence of an IMHB was suspected. For the matched pair in Figure 6.3, nearly identical EPSA values were measured and upon closer inspection of the topologies present in this example it became obvious that the control compound can form an IMHB between the N–H of the amide and the nonbridge N in the pyrimidine ring.

In conclusion, Goetz et al. reported a novel SFC-based chromatography method that enables identification of IMHBs through a pair-wise analysis. The majority of observed exceptions could be explained through the presence of additional topologies capable of forming IMHBs. These findings were supported by NMR data. The

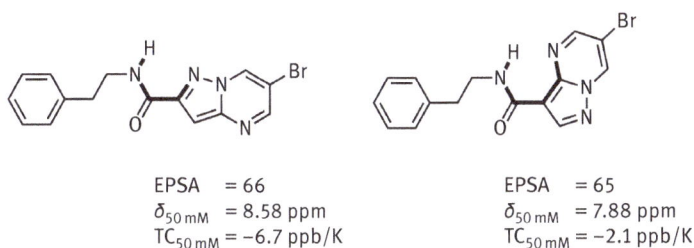

EPSA   = 66
$\delta_{50\,mM}$ = 8.58 ppm
$TC_{50\,mM}$ = −6.7 ppb/K

EPSA   = 65
$\delta_{50\,mM}$ = 7.88 ppm
$TC_{50\,mM}$ = −2.1 ppb/K

Figure 6.3: Example of a matched pair of compounds where an additional IMHBs was found in the control compound through EPSA measurements [19].

authors further concluded that EPSA has the potential to serve as an indicator for the presence of IMHBs and as a tool for indirect polarity measurement.

Encouraged by these findings it was envisioned that the methodology could also be extended to peptides. As highlighted before in this chapter peptides are commonly associated with poor bioavailability and as a matter of fact the majority of peptidic drugs are administered by injection [25]. Due to the structural nature of peptides a multitude of polar interactions can occur between both the backbone and the polar side chains of a peptide and the surface of a chromatographic stationary phase. These interactions contribute to the retention of the peptide on the column, which is further increased with higher exposure of polar groups.

Consequently, the chromatographic SFC method had to be adapted to the more polar analytes. In order to facilitate elution of the peptides the existing gradient of MeOH in supercritical CO_2 from 5% to 50% at 5% per minute was extended to reach 60% over the course of 11 min, and a 5 min hold period was added, increasing the total runtime from 12 to 17 min [26].

The EPSA values for a selection of cyclic peptides were plotted against Ralph Russ canine kidney cells (RRCK) permeability values, a subclone of the parental MDCK cell line. The plot revealed that all permeable peptides but one had an EPSA value of 80 or lower. In addition, a number of known cyclic peptides such as cyclosporine A, oxytocin, and a Lokey-type cyclic hexapeptide were evaluated, corroborating the hypothesis that low EPSA values and increased permeability are closely correlated [27]. Cyclosporine and the Lokey hexapeptide were found to have EPSA values of 71 and 78, respectively, with bioavailabilities in rate of 43.7% and 28.1%. On the other hand, oxytocin exhibits an oral availability of 0.85% and an EPSA value of 144. A statistical analysis of the RRCK permeability of 814 cyclic peptides revealed that compounds with an EPSA value of 80 or lower have a high likelihood to be either moderately permeable (34%) or permeable (38%), while cyclic peptides with an EPSA value of >100 generally do not exhibit significant passive membrane permeability.

The authors furthermore found the assay to be responsive to small changes in the molecular structure of peptides such as N-methylation, amino acid inversion, and substitution of single amino acids in the sequence. This feature, in combination with directed N-methylation of each amino acid (methyl walk) allows for the identification of IMHBs in cyclic peptides through monitoring EPSA values.

In conclusion, the EPSA methodology was established as a high-throughput chromatographic SFC-based tool to monitor polarity in peptides and drive the design of the latter toward increased permeability. Furthermore, EPSA is indicative of IMHBs in cyclic peptides. Besides medicinal chemistry teams at Pfizer, the methodology has found wide acceptance and use in the industry. Peptide chemists at Novartis in Switzerland have used this SFC chromatographic method to assess changes in the polarity of cyclic peptides upon N-methylation and the introduction of hydrophobic shielding [28].

Incorporation of aminobutyric acid (Abu) into the sequence of a cyclic hexapeptide with the intention to shield polar atoms resulted in high PAMPA permeability. Furthermore, the shortest SFC retention time (2.7 min) within a series of cyclic hexapeptides was found for the Abu derivative. These findings are in agreement with a high oral bioavailability of 39% in mice [28]. It was found that the SFC retention times obtained from the EPSA assay generally correlate well with PAMPA permeability and oral bioavailability in mice and rats.

References

[1] (a) E. Valeur, S. M. Guéret, H. Adihou, R. Gopalakrishnan, M. Lemurell, H. Waldmann, T. N. Grossmann, A. T. Plowright, Angew. Chem. Int. Ed. 56 (2017) 10294–10323; (b) M. Vieth, J. J. Sutherland, J. Med. Chem. 49 (2006) 3451–3453; (c) G. V. Paolini, R. H. B. Shapland, W. P. van Hoorn, J. S. Mason, A. L. Hopkins, Nat. Biotechnol. 24 (2006) 805–815.
[2] M. Toure, C. M. Crews, Angew. Chem. Int. Ed. 55 (2016) 1966–1973.
[3] (a) A. Alex, D. S. Millan, M. Perez, F. Wakenhut, G. A. Whitlock, Med. Chem. Commun. 2 (2011) 669–674; (b) N. Terrett, Med. Chem. Commun. 4 (2013) 474–475.
[4] B. Over, P. McCarren, P. Artursson, M. Foley, F. Giordanetto, G. Grönberg, C. Hilgendorf, M. D. Lee, IV, P. Matsson, G. Muncipinto, M. Pellisson, M. W. D. Perry, R. Svensson, J. R. Duvall, J. Kihlberg, J. Med. Chem. 57 (2014), 2746–2754.
[5] D. J. Craik, D. P. Fairlie, S. Liras, D. Price, Chem. Biol. Drug Des. 81 (2013) 136–147.
[6] (a) J. F. Woodley, Crit. Rev. Ther. Drug Carrier Syst. 11 (1994) 61–95; (b) N. Salamat-Miller, T. P. Johnston, Int. J. Pharm. 294 (2005) 201–216.
[7] (a) R. J. Ptachcinski, G. J. Burckart, R. Venkataramanan, Drug Intell. Clin. Pharm. 19 (1985) 90–100; (b) A. Fahr, Clin. Pharmacokinet. 24 (1993) 472–495.
[8] N. El Tayar, A. E. Mark, P. Vallat, R. M. Brunne, B. Testa, W. F. van Gunsteren, J. Med. Chem. 36 (1993) 3757–3764.
[9] (a) T. Rezai, J. E. Bock, M. V. Zhou, C. Kalyanaraman, R. S. Lokey, M. P. Jacobsen, J. Am. Chem. Soc. 128 (2006) 14073–14080; (b) T. R. White, C. M. Renzelman, A. C. Rand, T. Rezai, C. M. McEwen, V. M. Gelev, R. A. Turner, R. G. Linington, S. S. F. Leung, A. S. Kalgutkar, J. N. Bauman, Y. Zhang, S. Liras, D. A. Price, A. M. Mathiowetz, M. P Jacobsen, R. S. Lokey, Nat. Chem. Biol. 7 (2011) 810–817; (c) A. C. Rand, S. S. F. Leung, H. Eng, C. J. Rotter, R. Sharma, A. S. Kalgutkar, Y. Zhang, M. V. Varma, K. A. Farley, B. Khunte, C. Limberakis, D. A. Price, S. Liras, A. M. Mathiowetz, M. P. Jacobson, R. S. Lokey, Med. Chem. Commun. 3 (2012) 1282–1289.

[10] (a) S. Hess, O. Ovadia, D. E. Shalev, H. Senderovich, B. Qadri, T. Yehezkel, Y. Salitra, T. Sheynis, R. Jelinek, C. Gilon, A. Hoffman, J. Med. Chem. 50 (2007) 6201–6211; (b) S. Hess, Y. Linde, O. Ovadia, E. Safrai, D. E. Shalev, A. Swed, E. Halbfinger, T. Lapidot, I. Winkler, Y. Gabinet, A. Faier, D. Yarden, Z. Xiang, F. P. Portillo, C. Haskell-Luevano, C. Gilon, A. Hoffman, J. Med. Chem. 51 (2008) 1026–1034; (c) T. Rezai, B. Yu, G. Millhauser, M. P. Jacobson, R. S. Lokey, J. Am. Chem. Soc. 128 (2006) 2510–2511; (d) O. Ovadia, S. Greenberg, J. Chatterjee, B. Laufer, F. Opperer, H. Kessler, C. Gilon, A. Hoffman, Mol. Pharm. 8 (2011) 479–487.

[11] (a) P. Artursson, J. Pharm. Sci. 79 (1990) 476–482; (b) P. Artursson, C. Magnusson, J. Pharm. Sci. 79 (1990) 595–600; (c) L. Di, C. Whitney-Pickett, J. P. Umland, H. Zhang, X. Zhang, D. F. Gebhard, Y. Lai, J. J. Federico, III, R. E. Davidson, R. Smith, E. L. Reyner, C. Lee, B. Feng, C. Rotter, M. V. Varma, S. Kempshall, K. Fenner, A. F. El-Kattan, T. E. Liston, M. D. Troutman, J. Pharm. Sci. 100 (2011) 4974–4985; (d) A. Avdeef, Expert Opin. Drug Metab. Toxicol. 1 (2005) 325–342; (e) A. Avdeef, P. Artursson, S. Neuhoff, L. Lazorova, S. Gråsjö, S. Tavelin, Eur. J. Pharm. Sci. 24 (2005) 333–349.

[12] G. H. Goetz, L. Philippe, M. J. Shapiro, ACS Med. Chem. Lett. 5 (2014) 1167–1172.

[13] (a) R. E. Hubbard, M. Kamran Haider, Hydrogen Bonds in Proteins: Role and Strength, John Wiley & Sons, Ltd., Hoboken, NJ, 2001; (b) B. Kuhn, P. Mohr, M. Stahl, J. Med. Chem. 53 (2010) 2601–2611.

[14] (a) G. Lessene, B. J. Smith, R. W. Gable, J. B. Baell, J. Org. Chem. 74 (2009) 6511–6525; (b) C. J. E. Haynes, N. Busschaert, I. L. Kirby, J. Herniman, M. E. Light, N. J. Wells, I. Marques, V. Felix, P. A. Gale, Org. Biomol. Chem. 12 (2014) 62–72; (c) T. Cierpicki, J. Otlewski, J. Biomol. NMR 21 (2001) 249–261.

[15] M. J. Winningham, D. Y. Sogah, J. Am. Chem. Soc. 116 (1994) 11173–11174.

[16] M. Shalaeva, G. Caron, Y. A. Abramov, T. N. O'Connell, M. S. Plummer, G. Yalamanchi, K. A. Farley, G. H. Goetz, L. Philippe, M. J. Shapiro, J. Med. Chem. 56 (2013) 4870–4879.

[17] C. R. Clark, M. J. M. Wells, R. T. Sansom, J. L. Humerick, W. B. Brown, B. J. Commander, J. Chromatogr. Sci. 22 (1984) 75–79.

[18] (a) K. Valko, C. My Du, C. D. Bevan, D. P. Reynolds, M. H. Abraham, J. Pharm. Sci. 89 (2000) 1085–1096; (b) M. Kansy, F. Senner, K. Gubernator, J. Med. Chem. 41 (1998) 1007–1010; (c) S. Ong, H. Liu, X. Qui, G. Bhat, C. Pidgeon, Anal. Chem. 67 (1995) 755–762; (d) C. Pidgeon, S. Ong, H. Liu, X. Qui, M. Pidgeon, A. H. Dantzig, J. Munroe, W. J. Hornback, J. S. Kasher, L. Glunz, T. Szczerba, J. Med. Chem. 38 (1995) 590–594; (e) G. W. Caldwell, J. A. Masucci, M. Evangelisto, R. White, J. Chromatogr. A 800 (1998) 161–169.

[19] G. H. Goetz, W. Farrell, M. Shalaeva, S. Sciabola, D. Anderson, J. Yan, L. Philippe, M. J. Shapiro, J. Med. Chem. 57 (2014) 2920–2929.

[20] D. L. Goldfarb, D. P. Fernández, H. R. Corti, Fluid Phase Equilib. 158 (1999) 1011–1019.

[21] (a) M. T. Combs, M. Ashraf-Khorassani, L. T. Taylor, J. Chromat. A 785 (1997) 85–100; (b) L. T. Taylor, J. Supercrit. Fluids 47 (2009) 566–573; (c) E. Lesellier, C. West, J. Chromat. A 1382 (2015) 2–46.

[22] B. Kuhn, P. Mohr, M. Stahl, J. Med. Chem. 53 (2010) 2601–2611.

[23] L. L. Parker, A. R. Houk, J. H. Jensen, J. Am. Chem. Soc. 128 (2006) 9863–9872.

[24] T. Cierpicki, I. Zhukov, R. A. Byrd, J. Otlewski, J. Magn. Reson. 157 (2002) 178–180.

[25] D. J. Craik, D. P. Fairlie, S. Liras, D. Price, Chem. Biol. Drug Des. 81 (2013) 136–147.

[26] G. H. Goetz, L. Philippe, M. J. Shapiro, *ACS Med. Chem. Lett.* 10 (2014) 1167–1172.

[27] T. R. White, C. M. Renzelman, A. C. Rand, T. Rezai, C. M. McEwen, V. M. Gelev, R. A. Turner, R. G. Linington, S. S. Leung, A. S. Kalgutkar, J. N. Bauman, Y. Zhang, S. Liras, D. A. Price, A. M. Mathiowetz, M. P. Jacobson, R. S. Lokey, Nat. Chem. Biol. 7 (2011) 810–817.

[28] T. Vorherr, I. Lewis, J. Berghausen, S. Desrayaud, M. Schaefer, Int. J. Pept. Res. Ther. (2017). https://doi.org/10.1007/s10989-017-9590-8.

Eric Lesellier and Caroline West

7 Applications of supercritical fluid chromatography: Natural products in pharmaceutical, cosmetic, and food applications

Abstract: A thorough knowledge of the composition of natural products is desired in many application fields, for instance: for pharmaceutic and cosmetic applications where the identity of active species must be determined, or in traditional medicine and foodomics where the quality must be strictly controlled. This chapter describes the analysis of natural products with packed column supercritical fluid chromatography. The literature is reviewed, focusing on different aspects of method development that must be addressed when dealing with such samples: the selection of stationary phase depending on analyte polarity and structural features, the selection of mobile phase to elute polar and/or non-polar species, and fine-tuning operating parameters as temperature and pressure. The choice of detectors, preparative separations and two-dimensional systems are also briefly discussed.

Keywords: Secondary metabolites, plants, method development

7.1 Introduction

For 30 years, the use of packed column supercritical fluid chromatography (pSFC) has been rising up in numerous analytical fields because of many physical and chemical advantages of the fluids used as a mobile phase, in combination with a large range of packed columns available. Conversely and beyond the works achieved by pioneers, fundamental studies allowed for improved understanding of subtle interactions developed between the analytes, the mobile phase, and the stationary phases (SP), yielding successful separations in short analysis duration [1–5].

It is now well admitted that due to the low fluid viscosity that is around 10 times less than the one of liquids, pSFC can be performed with small particle size and high flow rate that greatly reduce the analysis time. Moreover, this increase in flow rate causes small reduction in the theoretical plate number because of the high diffusion coefficients of analytes in supercritical fluids, and can be achieved with only a modest

Eric Lesellier and Caroline West, University of Orléans, Institute of Organic and Analytical Chemistry, CNRS UMR 7311, France

https://doi.org/10.1515/9783110500776-007

increase in the pressure drop. These properties allow working with classical inlet pressure, that is, about 30–40 MPa despite the backpressure required to maintain the fluid in its super- or subcritical state.

pSFC is often compared to normal-phase liquid chromatography (LC). This comparison is due to the nature of carbon dioxide that has very low polarity, which induces the use of a polar SP to counterbalance the interactions. However, for a long time, reversed-phase liquid chromatography (RPLC) has been the most widely used technique in the field of natural products as in others. It can be explained by the fact that normal-phase LC, which uses toxic solvents, is unable to separate compounds that only differ by hydrocarbonaceous volume, that is, by the number of methylene (CH_2) or methyl groups (CH_3), whereas reversed phase also allows the resolution of compounds having different polar groups. This is also true in the case of pSFC. Moreover, some compounds found in natural products are rather non-polar and do not interact with polar SPs. Hydrophobic species are best analyzed with C18-bonded SPs to improve their retention and resolution. When a nonpolar C18 SP is employed, SFC could appear as an unidentified flying object in comparison to other chromatographic methods, because both phases are nonpolar. However, this combination does work, meaning that pSFC is a universal method of separation and able to perform separations with all types of SPs in packed columns, covering the normal and the reversed-phase areas of high-performance liquid chromatography (HPLC), and sometimes even the HILIC area [6].

The chemical structure of molecules in plants is varied due to the numerous functions that they are intended to perform in living organisms. Primary metabolites are the compounds that are necessary for the plant growth, for instance, sugars, lipids in seeds (triacylglycerols, TAGs), amino acids, or pigments that absorb light energy for photosynthesis. However, many stresses act on plants and compel them to synthesize secondary metabolites to either repel predators or resist extreme climatic conditions.

Secondary metabolites present various biochemical activities, which may be useful for human health and well-being in pharmaceutical or cosmetic products. However, these compounds can also display cell toxicity. Due to their biochemical properties, these compounds are also called "specific metabolites," and their identification becomes a major challenge. They can be divided into three main classes: polyphenols (flavonoids, lignans, phenolic acids), terpenes (mon-, di-, triterpenes, phytosterols), and alkaloids, with other smaller families such as vitamins, pigments, or tannins. For some of these compounds (e.g., flavonoids and triterpenes), two forms can be present in the plant: aglycone and glycoside, and the former is less toxic and used for storage in the plant. Additionally, in most families of natural products, there are structural and configurational isomers that require very high-separation performances and a well-suited choice of SPs. These metabolites exist in different parts of the plants: roots, rhizomes, leaves, barks, stems, seeds, or flowers.

Various extraction methods, from maceration and pressurized fluid extraction to classical distillation, are described in the literature. Initially mostly applied to the extraction of essential oil [7], the use of supercritical fluid extraction is gradually expanded to extract nonvolatile compounds. This rising interest may be due to the realization of the safety attributes of the fluid and its versatility in terms of solvating power, which depends on the temperature, the pressure, and the cosolvent added (most often ethanol) [8]. Numerous studies on extraction have shown that there are very few limitations to the solubility of most natural compounds in supercritical fluids. This, in turn, implies that these molecules can also be analyzed by SFC, although a large portion of supercritical fluid extracts are still analyzed by GC or HPLC.

This chapter will describe some fundamental points applied to the separation of natural compounds in relation to their polarity and structural differences. The separation issues related to natural products will be addressed reviewing (i) the classical retention behaviors related to the SP nature, (ii) the mobile phase composition, and (iii) the effects of temperature and backpressure changes.

7.2 The choice of Stationary phase

The selection of an adequate SP remains a major issue when developing a method with pSFC for the separation of natural products. It was shown several times that all types of SPs could be compatible with the CO_2-based mobile phase, from the nonpolar C18 to the polar silica. Determining the choice of SP chemistry should be the first step in method development to warrant a successful method. Obviously, carbon dioxide is a nonpolar fluid and is unable to dissolve polar compounds. The addition of a cosolvent (or modifier), and sometimes acidic or basic additives, is sufficient to improve the solubility of most analytes, but requires mobile phase optimization, as will be detailed later in the chapter.

Several papers reported the separation of the four tocopherol (vitamin E) and/or the four tocotrienols (structures of molecules shown in Figure 7.1). The retention order when using a C18 SP (L-column octadecylsiloxane [ODS]) was δ, γ, β, and α-tocopherols [9], whereas with a diol phase (nucleosil 100–50H), the retention order was changed to α, β, δ, and γ [10]. It shows that the retention on the C18 phase was ruled by the number of methyl groups (increasing hydrocarbon volume causes an increase in retention on the nonpolar SP), while the retention on the diol phase depended on the steric hindrance of the hydroxyl group (on carbon 6) caused by methyl groups in positions 5 and 7 of the chromanol head.

In addition, the retention of the four tocotrienols, which have the same skeleton as tocopherols but with three additional double bonds in the phytyl chain, was improved on the polar (diol) SP, and the overall separation of the eight compounds was achieved in SFC in 40 min with methyl-*tert*-butyl-ether as the cosolvent.

Recently, a rapid screening of six varied SP resulted in the selection of an amino phase (Luna NH$_2$) to achieve the separation of seven of the eight vitamin E congeners in 4 min with a linear gradient elution program (CO$_2$–ethanol containing 0.1% formic acid, from 3.5% to 8%). The retention order was α, β, γ, and δ-tocopherols (Figure 7.1) [11]. This example shows the strong improvement in the analytical performance afforded by a column packed with smaller particles (1.7 instead of 5 μm) and smaller internal diameter

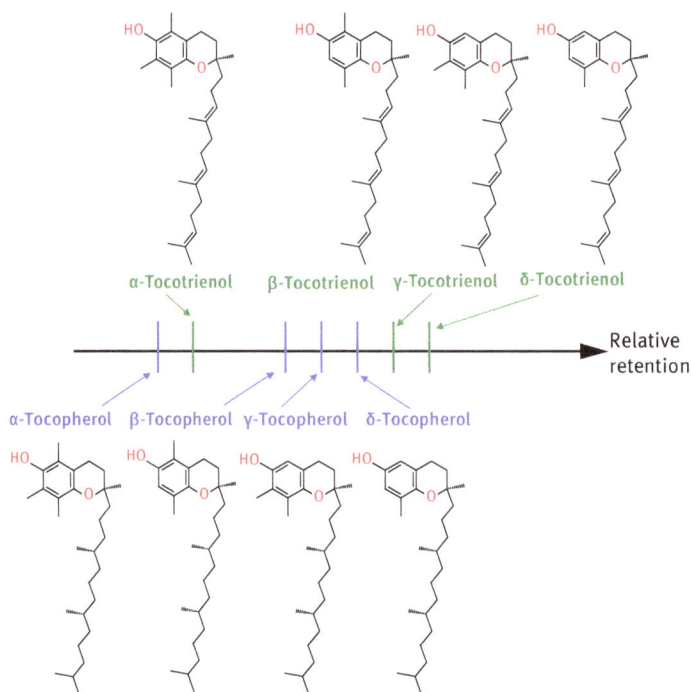

Figure 7.1: Retention order observed for tocopherols and tocotrienols on an aminopropyl-bonded SP with carbon dioxide–ethanol gradient [11].

(2 instead of 4.6 mm). The same retention order for the four tocopherols was reported when using neat CO_2 at high pressure (32 MPa) with a polyethylene glycol SP [12].

Carotenoid pigments, which play many different roles in plants [13], are often separated with ODS-bonded phases due to their high hydrophobicity [14–16]. Among the varied ODS phases available, monomeric phases with high bonding density or polymeric phases favor the separation of the numerous *cis/trans* isomers of carotenes [17, 18]. Monomeric phases with high bonding density also favor the separation of neoxanthin and violaxanthin (structural isomers) [16]. The separation of chlorophylls a and a′ (which only differ from the R,S configuration of C_{21}) and their transformation products (pheophytin and pyrochlorophyll) was achieved with a Spherisorb ODS 2 column [19]. However, the two isomers of the more polar chlorophyll b were unresolved.

A nice tandem-column system comprising a Sunfire C18 and a Viridis 2-Ethylpyridine (2-EP; Waters) was used to separate eight carotenoids and xanthophylls (astaxanthin, β-carotene, canthaxanthin, echinenone, lutein, neoxanthin, violaxanthin, and zeaxanthin) extracted from microalgae [20]. These two columns also avoided the coelution of carotenoids with chlorophyll pigments (six additional peaks). Such column coupling is favored in SFC due to the low fluid viscosity. Another example of the tandem-column system was proposed for the separation of azadichrachtin a and b, nimbin, and salannin, which are four triterpenoids from the seeds of *Azadirachta indica* A. Juss. and have insect-repellent and growth disrupting properties. Their separation was obtained by coupling a Kromasil NH$_2$ (Eka-Nobel) and a Cyano (YMC) column with 10% methanol as a cosolvent [21]. Another example was the coupling of a Symmetry C18 (Waters) and a Zorbax SB-C18 (Agilent) for the separation of hop extracts (xanthohumol) [22].

While coupling different columns takes advantage of complementary selectivities, coupling identical SPs is also of interest to improve efficiency, thereby improving resolution between critical pairs of peaks despite the greater pressure gradient inside the column [16]. Three ODS columns (Inertsil) were coupled for the separation of polyprenols, which are linear polymers of isoprenoid units (octa- or nonadecaprenol: C$_{90}$H$_{145}$OH and C$_{95}$H$_{153}$OH with 18 or 19 double bonds, respectively) from *Eucommia ulmoides* [23]. The elution of these highly hydrophobic compounds that have a very high molecular weight required a high amount of cosolvent (ethanol) and a high temperature (130 °C). Other studies reported the use of coupled ODS phases, up to 125 cm, to achieve a complete separation of TAGs from vegetable oils, including the pairs of positional isomers (LLLn/OLnLn; LLL/OLLn; and OLL/OOL) [24, 25].

For several years, superficially porous particles, a modern particle technology, have favored the use of longer columns. These particles have a solid nonporous core in which neither the mobile phase nor the analytes may enter, and a porous shell covering the solid core, in which the equilibrium takes place. In a sense, the bottom of the porous part of the particles is not as deep compared to fully porous particles, which reduces the band dispersion. The resulting thinner peaks yield better separation performances. Moreover, because of the decrease in the phase porosity, lower inlet pressures are required to ensure identical flow rates. The behavior of these superficially porous particles in SFC was extensively studied [26]. This last point, combined with the low viscosity of supercritical fluids, allows the use of very long columns (75 cm length). More than 120,000 effective theoretical plates were achieved with an inlet pressure not exceeding 35 MPa. This ultrahigh-performance method was used for the separation of TAGs from edible oils [27] or oils extracted from seeds, for instance, *Kniphofia uvaria* that also contains anthraquinones [28, 29]. Figure 7.2 shows the impressive separation of TAGs from rapeseed oil, including the pairs of positional isomers (as mentioned earlier). The authors described the method as ultrahigh-efficiency low-pressure SFC [27].

In recent studies, the selection of the SP was most often done from two sets of columns, which are related to the SFC system manufacturer.

With the Agilent Aurora SFC system, Zorbax columns were mainly employed. For instance, SB-C18 and RX-Sil were used for the separation of 12 flavonoids and the quantification of kaempferol, luteolin, quercetin, luteoloside, and budleoside in *Chrysanthemum morifolium* Ramat [30]. Zorbax SB-C18, Eclipse Plus-C18, Eclipse XDB-C18, SB-CN, and RX-Sil were tested for the separation of colored compounds in paprika Oleoresins [31]. The C18 phases were selected for the unsaponified extracts, whereas a better separation of the saponified samples was obtained with the silica phase RX-Sil. The latter was also selected for the separation of nine triterpenoid saponins [32].

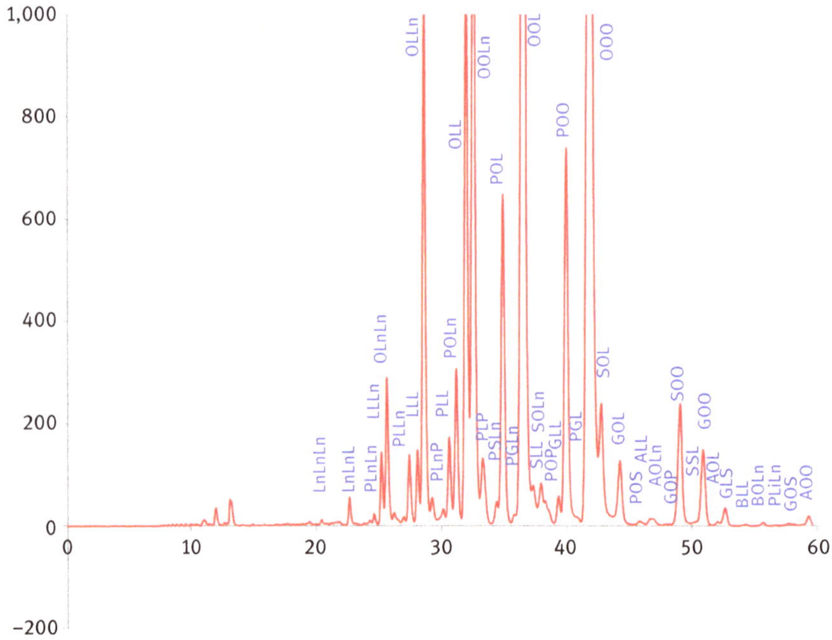

Figure 7.2: Analysis of rapeseed oil TAGs with 75 cm of columns packed with superficially porous particles (60 cm Kinetex C18 and 15 cm Accucore C18, 4.6 mm i.d., 2.7 µm). Operating conditions: carbon dioxide–acetonitrile–methanol 88:10.8:1.2 (v/v/v), 10 MPA, 17 °C, 1.6 mL/min, UV detection 210 nm [27].

With the Waters Acquity UPC^2 system, BEH (hybrid silica), BEH-2EP (2-EP), CSH Fluoro-phenyl, HSS SB C18 phases and the most recently released Torus 1-AA (amino-anthracene), Torus 2-PIC (2-picolylamine), Torus DEA (diethylamine), and Torus Diol are often selected because of the diversity or the ligands providing a range of complementary selectivities, and for their small particle size (sub-2 µm fully porous particles for all of them) ensuring high efficiency.

After the first screening step, the hydrophobic 1-AA phase (containing an anthracenyl ligand) was retained for the separation of 10 carotenoid pigments in microalgae [33]. However, this 1-AA phase displayed strongly tailing peaks for the analysis of furostanol saponins (steroidal saponins), and thus, the diol phase was preferred in this case [34]. The diol phase was also retained for the analysis of phenolic compounds from grapeseed extracts, successfully separating catechin from epicatechin, but failing to separate protocatechuic and caffeic acids [35].

The BEH phase (hybrid silica) was selected for the separation of six kavalactones (*Piper methysticum*), which display anticancer and antihyperglycemic properties [36], and also for the analysis of nine isoflavones, five aglycones, and four glycosylated [37].

The 2-EP SP has long been a favorite among pSFC users, particularly in the field of pharmaceutical analysis. The BEH 2-EP was used for the analysis of tri-and diacylglycerols in numerous vegetable oils to investigate edible oil adulterations [38]. It was also selected for the analysis of nine derivatives of benzoic and cinnamic acids, which are present in many fruits and berries and possess diverse biological activities [39]. The same SP was employed for the analysis of four anthraquinones from seed oil (*Kniphofia uvaria*) [28]. This SP also provided an efficient separation of annonacin and squamocin, which are polyketides from *Annonacae* family [40]. It was also selected for the separation of carnosic acid and carnosol, with a greater retention of the acid versus the alcohol, whereas the alcohol compound was more retained than the acidic one with a silica phase [41]. Other phenolic compounds (o-vanillin, styracin, vanillin, trans-cinnamic acid, vanillic acid, and shikimic acid) from *Liquidanbaris* resins were also perfectly separated with the BEH 2-EP [42].

Twelve isoflavones (five aglycones and seven glycosides) were separated with the Torus DEA [43], whereas the CSH fluorophenyl and the HSS C18 SB were preferred for seven other flavonoids [44]. The glycosides forms were more retained than the aglycone ones in any of the SPs. Moreover, the retention order between the CSH fluorophenyl and the HSS C18 SB was identical for the seven studied compounds. Such behavior was also reported for the separation of positional isomers of decursin and decursinol angelate, two pyranocoumarins from *Angelica gigas* Nakai with the BEH, BEH 2-EP, HSS C18, and CSH [45], and for the separation of Kaurenoic and contimentalic acids (diterpenoic acids) from *Aralia contimentalis* Kitagawa, with the same four Acquity phases plus the four Torus ones [46]. In the last two cases, the separation of the pairs of isomers was not mainly related to the SP chemistry but to the conformational structure of isomers, which induces different contact surface area between the compounds and the adsorbent.

Several studies concluded that the Acquity HSS C18 SB achieved higher performance compared to the other Acquity phases. This C18 phase was able to achieve the separation of the five main coumarins in *Ammi visnaga* fruits, mainly the one between Dihydrosamidin and Visnadin, the two positional isomers of the mixture [47]. It was also successfully used for the analysis of biogenic amines (precursor for the synthesis of hormones) in fermented foods [48], and for the impressive separation of γ and α C18:3 free fatty acids [49]. The separation of Emodin and Aloe-emodin (anthraquinones) from aerial part of rhubarb was also achieved with the same SP [50].

Among the other SPs used for SFC, some most significant ones should also be cited. For instance, a C30 phase successfully separated apocarotenoids (capsorubinal), but failed to separate apocarotenoid fatty esters [51]. A cyanopropyl-bonded phase also provided some interesting separations, for instance, taxicin I and II, which are taxane anticancer drugs isolated from the bark of *Taxus brevifolia* [52].

On a more fundamental aspect, numerous publications have reported the chromatographic behavior of polar, nonpolar, aromatic, and penta-fluoro phenyl SPs on the basis of the modified solvation parameter model, describing seven interaction types: (*e*) charge transfer, (*s*) dipole–dipole, (*a*) hydrogen bond donor (acidic) and

(b) acceptor (basic), (v) dispersion (London), and ionic interactions with anions (d^-) and cation (d^+) [6]. A unified classification using a spider diagram allows to compare more than 80 SPs and can be used as a guide for the column selection or the better understanding of the observed chromatographic behaviors [6, 53–60].

Figure 7.3 displays the location of SPs with recent particle types (sub-2 μm fully porous and superficially porous sub-3 μm) on this classification [61]. For instance, among the polar SPs, silica, amino, cyano, and 2-EP display a strong basic character, which is suited to the retention and separation of acidic compounds, whereas the HSS C18 SB displays the behavior of a polar-embedded or polar-endcapped phase, which explains why the retention order often observed on this phase is similar to that obtained on polar SPs.

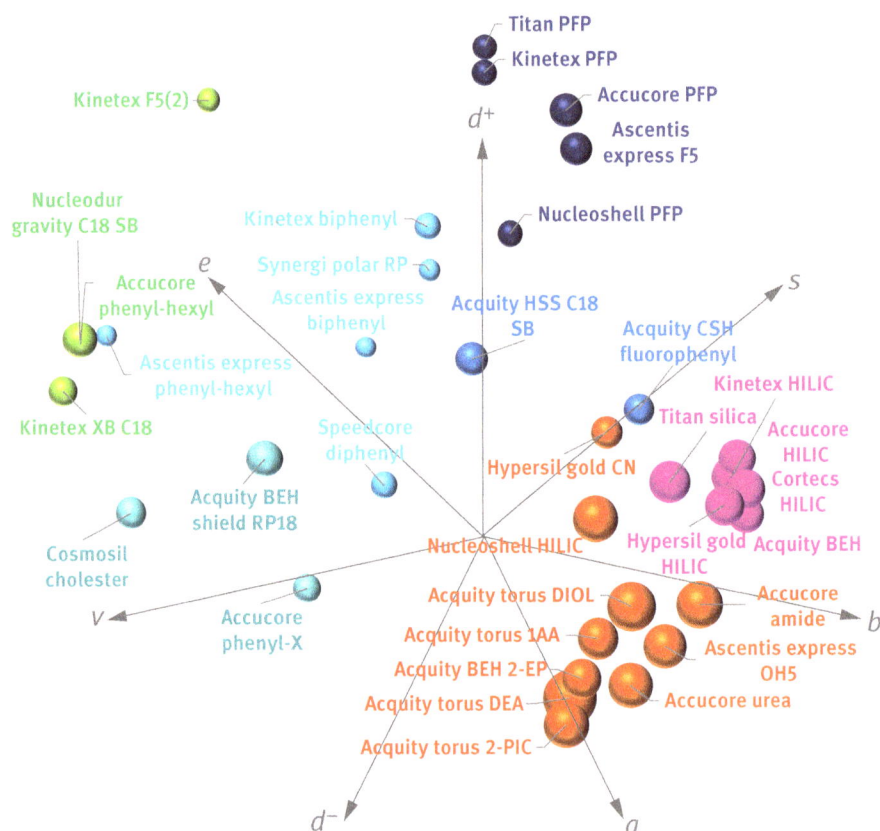

Figure 7.3: Classification of SPs based on the modified solvation parameter model calculated with seven descriptors for 35 columns with sub-2 μm fully porous or sub-3 μm superficially porous particles. Columns are identified with colors according to their similarity in selectivity. Chromatographic conditions: CO_2–MeOH 90:10 (v/v), 25 °C, 150 bar, 1 or 3 mL/min depending on column dimensions. Updated from [61].

On the basis of this classification, a large screening of columns was achieved for the separation of furocoumarins in citrus oil [62]. These compounds display phototoxic properties and their presence in cosmetic and fragrance products is strictly controlled. Using a Discovery HS F5 pentafluorophenyl SP, 16 identified psoralen derivatives (substituted in position 5 and 8) and around 10 additional compounds were separated in less than 12 min. An impressive resolution between the xanthotoxin and bergapten was achieved. These compounds are isomers: they differ by the position of one methoxyl group in position 8 and 5, respectively. The CSH Fluoro-phenyl phase was also retained for the analysis of seven coumarins and furocoumarins from the roots of *Angelica dahurica* [63]. This wide-selectivity screening approach was also successfully applied to the analysis of three classes of triterpenoids. It allowed the separation of numerous isomers of monoalcohol, diol, and acidic triterpenes (ursolic, oleanolic, and betulinic) with a Synergi Polar RP SP (see location on the spider diagram in Figure 7.3), allowing for the comparison of the composition of peels of different apple varieties [64]. Figure 7.4 shows the perfect separation of α- and β-amyrin, two positional isomers of monoalchohol triterpenes, obtained with an Alltima HP C18 HL phase. Porous graphitic carbon (Hypercarb) is also an interesting SP for the separation of isomers of monoterpenes in essential oils, which are volatile compounds. The separation of nerol, geraniol, and linalool was obtained with 10% of modifier in 2 min [65].

The ultimate challenge for the chromatographer is the separation of compounds having asymmetric carbons in their skeleton. These separations require the use of enantioselective SPs for enantiomers, whereas diastereomers can be separated with achiral phases.

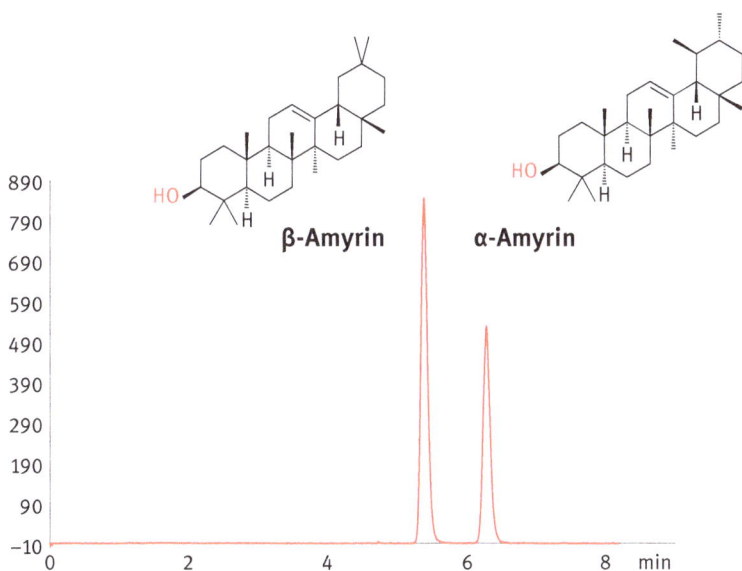

Figure 7.4: Separation of two isomers of triterpenoids, α- and β-amyrin. Alltima HP C18 HL (250 × 4.6 mm, 5 μm), carbon dioxide–methanol 90:10 (v/v), 25 °C, 15 MPa, ELSD.

Vitamin K1 has three chiral centers in its isoprenoid tail leading to the potential presence of eight stereoisomers. Using an amylose-based chiral SP (RegisPack), the separation of seven peaks from a commercial standard was obtained with 5% of methanol in carbon dioxide (isocratic elution) [66]. The enantioseparation of γ-lactones (natural flavors) was studied with four chiral selectors (Chiralpak AD, Chiracel OD, OB, and OJ) and four cosolvents (methanol, ethanol, isopropanol, and acetonitrile) [67]. The separation of seven racemic γ-lactones (γ-C6 to γ-C12) was achieved with Chiralpak AD (amylose *tris*-3,5-dimethylphenylcarbamate), the best suited phase, and with a low isopropanol percentage (from 1.5% to 3%). Spirocyclic norisoprenoid (theaspiranes, theaspirones, and vitispiranes), other flavor compounds present in various plants (roses, vanilla, passion fruits, tea, oak wood), contain two stereogenic centers resulting in four possible stereoisomers [68]. The Chiralpak IA and IF, in combination with acetonitrile as a cosolvent, afforded the best separations.

Rhizopine (4-amino-6-methoxycyclohexane-1,2,3,5-tetraol) is an opinelike compound present in root hairs of *Medicago sativa*, due to the infection of rhizobia bacteria [69]. The absolute configuration of natural rhizopine was studied by fraction collection of the peracylated derivatives separated with a Chiralpak AD-H and a methanol cosolvent with trifluoroacetic acid (TFA). The separation of four lignan diastereoisomers (galbelgin, ganschisandrin, galgravin, and veraguensin) from *Peper Kadsura* was studied with achiral (BEH, BEH 2-EP, HSS C18, CSH Fluoro-phenyl, and X-amide) and chiral phases (home-made polysaccharides of cellulose and amylose tris-3,5-dimethylphenylcarbamate) [70]. Although all achiral phases failed to separate the four compounds, the amylose enantioselective phase achieved the separation with 15% of methanol in 10 min. Seven pairs of 25R/S ergostane (tetracyclic triterpenoids) epimers form *Antrodia* camphorate were separated with the BEH 2-EP phase, whereas the Chiracel OJ-H phase was suited for the separation of each pair of 25/R/S epimer [71].

The separation of regioisomeric TAGs (position of alkyl chains on the three possible positions of TAG, *sn*-1, *sn*-2, or *sn*-3) in edible oil was successfully obtained with a C30-bonded SP, for TAG having saturated chain S or P (SSO/SOS, PPO/POP), whereas the separation of TAG having unsaturated ones was not achieved (LLO/LOL). To achieve this separation, CO_2/methanol (20–30%) and flow rate gradient (2–3 mL/min) were used [72].

7.3 Analytical conditions

7.3.1 Mobile phase composition

As mentioned in the introduction, the addition of organic solvents to carbon dioxide is required to enhance analyte solubility and improve analyte elution from the column.

Methanol, ethanol, isopropanol, and acetonitrile are most often used as the mobile phase. Acetonitrile is mainly suited for hydrophobic compounds with double bonds, for instance, for 17-polyacetylene (falcarinol and falcarindiol) [73]. Methanol/acetonitrile mixtures in varied proportions were selected for positional isomers of free fatty acids [49], TAGs [24, 25, 27], or coumarins [47]. Methanol favors the elution of alcohol and acidic compounds, whereas acetonitrile is sometimes required to improve the separation of compounds with double bonds [47, 49]. Ethanol is less toxic than methanol and was retained for its health and safety qualities to separate furocoumarins in citrus essential oil [62] or for the separation of acetogenins possessing antitumoral activity [40].

Isocratic elution mode with low proportion of cosolvent or small gradient slopes are generally retained for the separation of isomers or the less polar compounds (tocopherols, di- or triterpenes, furocoumarins, and TAGs). For instance, ethyl acetate at 2.5% allowed the separation of cis/trans isomers of vitamin K1 on a silica SP [74]. The percentage of a modifier typically varies from 0.5% to 40%, and the gradient elution mode is often used to reduce the duration of analysis. Due to numerous phenomena (e.g., mobile phase adsorption onto the SP surface, mobile phase density changes, and solvation variation in the cytobatic region of the analyte) [1–4], the retention ($\log k$) decrease with solvent proportion is usually not linear, regardless of the analyte. Besides, the solvent addition may also induce retention inversion that can modify the separation quality. For instance, the elution order of β- and γ-tocopherol was reversed around 10% methanol [9]. Similarly, the pair of TAGs POL/OLL reversed at 15% acetonitrile [24]. Understandably, alcohol cosolvents yield a greater retention variation for compounds that are capable of interacting with hydrogen bonds. For instance, the ethanol increase caused a greater retention change for byakangelin and oxypeucedanin hydrate (possessing two hydroxyl groups in the side chain) compared to more hydrophobic furocoumarins, leading to elution order changes [62].

Additives are also often added to the cosolvent to improve the solubility of polar analytes, to favor symmetric peak shapes, sometimes to modify the selectivity for critical pairs of peaks, and last but not the least to reduce secondary interactions with the SP during the solute elution (mainly with strong polar sites). Indeed, the absence of water in the mobile phase favors unwanted additional interactions between compounds and SP, possibly leading to broad or asymmetric peaks. Many different acidic additives were used (the numbers in parentheses indicate the percentage of additive in the cosolvent). Formic acid was used to separate diterpenoic acids (0.1%) [46], isoflavones (0.5%) [43], and tocopherols and tocotrienols (0.1%) [11]. Phosphoric acid was used for isoflavones (0.05%) [37] and flavonoids (0.1%) [30]. Among isoflavones, the addition of phosphoric acid was especially efficient to improve the peak shape of Genistein, which possesses three hydroxyl groups in positions R2, R4, and R5, and for the glycosylated isoflavones (Genistin, Glycitin, Daidzin, and Pueranin) [37], but without changes in retention and chromatographic selectivity, whereas formic acid also favored selectivities of couples of isoflavones [43]. TFA (0.01%) was used for phospholipids [75] and Sinalbin, a major glucosinolate in seeds of *Sinapis alba* (mustard) [76]. Methanesulfonic acid (1%) was used for ascorbigen, a degradation product produced in myrosinase-catalyzed hydrolysis of arylmethyl-glucosinolates [77], for which it was preferred over TFA to reduce the background absorption at low wavelength (217 nm) during the elution gradient. Similarly, citric acid (0.5%) was employed for phenolic compounds, for which it was found to be more efficient than TFA with a diol-bonded phase due to its acidic polyfunctionality [35]. Ascorbic acid at pH = 2 was recommended for the analysis of glucobrassicin derivatives and their oligomers [78]. Oxalic acid 10 mM was added to methanol to reduce the peak tailing of curcuminoids (curcumin, demethoxycurcumin, bisdemethoxycurcumin) extracted from the rhizome of *Curcuma longa* [79].

Naturally, these results were obtained with varied amounts of additives in cosolvent and different gradient slopes, meaning that the additives concentration in the mobile phase varied to a lesser or greater extend, and last but not the least, with different SPs. It is therefore difficult to draw any general conclusion on the applicability of each of these additives.

Basic additives are most often, but not limited, to be used for analyzing basic species. DEA or isopropylamine (IPA) were often selected, for instance, to elute lactones (0.6% DEA) [36] or coumarins (0.1% DEA), to favor the separation of xanthotoxin and imperatorin [63], or of aleomycin, a mixture of sulfomyocin and promothiocin peptides produced by fermentation of *Streptomyces arginensis* (0.2% IPA) [80]. The addition of triethylamine in the mobile phase was also suggested to improve the separation of alkaloids with an aminopropyl-bonded SP (e.g., narcotine, papaverine, thebaine, ethylmorphine, codeine, and morphine) and applied to the analysis of a poppy straw extract [81].

Mixtures of an acid and a base were sometimes advised in the past. For instance, a mixture of DEA and TFA (0.5% each in methanol) was reported to favor the so-called deactivation of a cyanopropyl phase, thereby improving peak shape and decreasing retention of atropine, an alkaloid from *Atropa belladonna* [82]. More recently, the use of a salt was often advocated. For instance, ammonium acetate (5 mM) was selected for the separation of nucleobases, nucleosides, and ginsenosides from Ginseng, mainly to improve the selectivity of the glycosylated ginsenoside [83]. In another paper, a solution of aqueous ammonia was added to methanol to inhibit a methoxylation reaction of furostanol saponins during the elution [34].

Water is also considered as an additive because of the low amount that can be added to CO_2. Water molecules adsorbed on the silanol groups onto the silica surface, leading to a retention decrease of alkaloids [81], while the effect was negligible on an amino phase. The addition of 5% water to methanol was beneficial for the separation of sesquiterpene lactones (from *Asteraceae* or *Umbelliferae*), which display different biological or therapeutic activities [84]. Varied amounts of water (from 0% to 12%) in methanol were tested with a silica column for the separation of triterpenoid saponins. The addition of 8–10% of water reduced the retention of compounds and improved both selectivity and peak shape [32]. It is well recognized that the mixture of water and CO_2 leads to the formation of carbonic acid [85]. Carbon dioxide saturated with water was used to elute and separate vitamins K, E, and D with a diol SP, but the stability of the mobile phase composition was limited to 2 h [86]. In another study [87], increasing methanol proportion in carbon dioxide reduced the retention time of sugars and polyols, while introducing water in methanol caused an increase in retention time. As often observed with numerous additives and/or modifiers, one of the assumptions is that water was strongly adsorbed onto the silica surface, favoring a partition mechanism between the mobile phase and the layers of adsorbed water. Besides, the addition of water favored the efficiency both on a nonpolar SP (tetramethylsilyl) and on a polar phase (silica). However, the presence of water also favored the mutarotation of sugars leading to the simultaneous presence of anomeric forms, inducing significant band broadening.

As appears in these examples, an additive may favor the separation through many different mechanisms. Another possibility is an ion-pairing mechanism. For example, di-octyl sodium sulfosuccinate was used as a counterion (100 mM) in methanol (15%) to elute the cationic berberine and palmitine from Cortex Phellodendri through ion pairing, allowing their solubility in the CO_2-based mobile phase [88].

Extensive retention behavior studies were carried out with carotenoid pigments and TAGs with C18-bonded SPs [25, 89–91]. The retention variations of these compounds versus the cosolvent percentage (from 5% to 40%) were not monotonous: first a retention decrease occurred when the modifier content was increased to 15–20%, then a retention increase was observed for higher percentage of the modifier. The same behavior was reported for TAG with pure CO_2 due to the temperature change, from 5 °C to 80 °C [92, 93]. It was shown that the minimum retention was related to the compound chemical structure (polarity), and would appear at different operating conditions (cosolvent percentage) for different compounds. As a result, changing the mobile phase composition did lead to retention inversion between some pairs of compounds. For instance, the position of minimum retention was lower for α-, β-, γ-carotene and lycopene (around 25% acetonitrile) in comparison to the more polar β-cryptoxanthin that possesses an additional hydroxyl group (30–35% acetonitrile) and lutein and zeaxanthin with two hydroxyl groups (40% acetonitrile) [1].

Because retention changes are different in relation to the chemical structure of the analytes, it is possible to use these chromatographic behaviors to help identifying the analytes. For example, retention patterns relating retention (log k) to partition number (which is equal to the total chain carbon number minus twice the chain double bond number) for TAGs are well known. For these TAGs, it was shown that the plotting retention changes (k_2/k_1) versus log k_1 when the modifier, pressure or temperature was changed could be helpful for their identification [24].

7.3.2 Temperature and backpressure

Due to the constraints of some instruments, most of the studies with regard to temperature are conducted at temperature values higher than 30 °C. However, an increase in the temperature most often increases the analysis duration due to the decrease in the fluid density, that is, in the eluting strength. Naturally, the extent of these retention changes and, consequently, selectivity variations are also related to the chemical structure of the compounds. For instance, from a study on furocoumarins, the greater retention variations were observed for byakangelicin and oxypeucedanin, where both have two hydroxyl groups on the side chain in position 8 and 5, respectively [62]. Lower temperatures favored the selectivity between two flavonoids, genistein and calycosin (40 °C) [44], between two carotenoid pigments, lutein and zeaxanthin (28 °C) [20], between two coumarins, khellin and visnagin (30 °C) [47], or between phenolic compounds, catechin and epicathechin (40 °C) [35]. A temperature of 20 °C was selected to improve the separation of triterpenoid saponins, mainly kudinoside C and G (tetraglycosylated compounds) [32] and also for triterpenes aglycones from apple peel [64]. For the analysis of the complete composition of seed oil, including free fatty acids, di- and triglycerides, a temperature of 9 °C was selected [94].

In other cases, the increase in temperature favored the separation between protocatechuic and caffeic acid (80 °C) [35] or between three flavonoid hyperosides, luteoloside, and myricetin (40 °C) [30].

Without detailing all the results from the literature, the change in backpressure was mostly negligible on retention and selectivities. Consequently, it would be advisable to maintain the backpressure at a high value to ensure a fluid density above 0.8 g/mL and a low fluid compressibility, and to improve method robustness.

7.4 Detection conditions

Many natural compounds (pigments, phenolic compounds, etc.) are colored or absorb light at specific wavelengths, explaining why UV/DAD detector is classically used in SFC with a pressure-resistant UV cell because of the pre-BPR location of the detector [4]. Other compounds (typically sugars, amino acids, or terpenes) display weak or no UV absorbance, and thus, open detectors should be used, like mass spectrometer (MS) or evaporative light-scattering detector (ELSD), which are both adequate for quantitation purposes. Table 7.1 presents the numerous studies achieved in SFC-MS [11, 30, 32–34, 38, 40, 49, 51, 72, 82, 94–104]. Coupling SFC and MS requires the presence of a polar solvent (often methanol), first to ensure the proper solvation of the compounds to avoid any precipitation in the transfer capillary located between the BPR and the MS source, and secondly to favor the ionization of the analytes [97]. The choice of electrospray ionization (ESI), atmospheric pressure chemical ionization (APCI), or atmospheric photoionization (APPI) is related to the ability of the compound to form ions with each type of ionization source. However, the APCI source is more capable of handling high flow rate of CO_2 eluting from the SFC system.

Obviously, MS is helpful in identifying both targeted and untargeted compounds. For instance, the carotenoid pigments of habanero pepper are oxidatively cleaved into

Table 7.1: Sample applications using SFC-MS.

Target analytes	UV detection	MS type	Ionisation // detection mode	Splitter position	Make-up fluid	LOQ	
Triterpenoid saponins	Weak/no UV absorbance	Single quadrupole	ESI + // SIM		MeOH/amonium acetate 10 mM 0.3 mL/min pre-BPR		[32]
TAGs	Weak/no UV absorbance	Q-ToF	ESI+ // m/z 300–1,200				[38]
Carotenoids	430 nm	Q-ToF	ESI+ //			10–50 ng/mL	[33]
Free fatty acids	Weak/no UV absorbance	Triple quadrupole	ESI– // SIM				[49]
Tocopherols/ tocotrienols	295 nm	Q-ToF	APPI // m/z 50–1,000		MeOH 0.2 mL/min pre-BPR	0.05–0.25 µg/L	[11]
Gingenosides	Weak/no UV absorbance	Single quadrupole	ESI+ // m/z 50–500/SIM	pre-BPR	MeOH/ammonium acetate 5 mM 0.3 mL/min pre-BPR		[30]
Atropine		Single quadrupole	APCI+ // SIM m/z 290 amu 30 V				[83]
Furostanol saponins		Q-ToF	ESI– //				[34]
Phycoecdysteroids			Thermospray chemical ionisation	Pre-BPR			[95]
Apocapsorubinals		Triple quadrupole	APCI +/– // SIM and multiple reaction monitoring				[51]
Artemisin		Single quadrupole	APCI+ // m/z 100–500 amu				[96]
TAGs	Weak/no UV absorbance	Single quadrupole	APCI+ // m/z 500–1,000 amu		MeOH 1.5 mL/min		[97]
Ginkgolic acids	305 nm	Single quadrupole	ESI– // m/z 300–500 amu	pre-BPR	MeOH/ 8 mM ammonium formate/0.5% FA; 0.6 mL/min	100 ng/mL	[98]

(continued)

Table 7.1: (continued)

Target analytes	UV detection	MS type	Ionisation // detection mode	Splitter position	Make-up fluid	LOQ	
Pyridine alkaloids	226 nm	Triple quadrupole	ESI+ // m/z 200–500 amu				[99]
Lipopolysaccharides		Triple quadrupole	APCI+ //			10 µg/g	[100]
Canabinoids	220 nm	Single quadrupole	ESI +/– // m/z 100–800 amu		MeOH/ 8 mM ammonium formate/0.5% FA	5–10 µg/mL	[101]
Lipids	Weak/no UV absorbance	Single quadrupole	ESI+ // m/z 300–1,200 amu		MeOH/ 0.1% ammonium formate 0.1 mL/min		[102]
Polyene phosphatidylcholine		Triple quadrupole	ESI+ //		MeOH/ 0.2% FA 0.2 mL/min	0.5–50 ng/mL	[103]
TAG, DGA, anthraquinones		Q-ToF	APCI+// m/z 50–1,400 amu		MeOH; IPA; MeOH/formic acid 0.1–0.5 mL/min		[94]
Acetogenins		Q-ToF	ESI+ // m/z 100–950 amu		MeOH/lithium 12 mM 200 µL/min	5–10 pmol inj	[40]
C17-polyacetylenes		Single quadrupole	APCI+ // m/z 200–1,400	Pre-BPR			[73]
Lignin derivatives		Q-ToF-MS	ESI– //		MeOH/5mM ammonia 0.2 mL/min	1–2 µg/mL	[104]

apo form. Due to the numerous double bonds onto the polyenic central chain, this cleavage could occur at different positions, which can be identified by the single-ion monitoring (SIM) with m/z values of the radical anion and by the multiple reaction monitoring experiments that provide quantifier (Q) and qualifier (q) transitions [51]. The identification of compounds from the mass of fragments can be complicated by in-source collision induction, as was reported for terpene lactone from *Ginkgo biloba* [98]. Comparison and selection of varied sources and ionization modes can be achieved by direct infusion [33]. However, in direct infusion, the compounds are usually dissolved in pure solvent, whereas when they are analyzed with supercritical fluids, they may be introduced in the MS source with a solvent proportion lower than 10%.

With ESI in positive ionization mode, the ginsenosides were detected as sodium adduct ions ($[M + Na]^+$) [30]. In that case, the better additive added to methanol in the mobile phase was not the same with regard to the chromatographic (ammonium acetate) or MS (formic acid) performances. Formic acid in the mobile phase was also selected for triperpenoid saponins as the best compromise between the MS signal intensity and overall resolution, and the addition of ammonium acetate with a post-column make up increased again the MS signal for three of the nine kudinosides [32]. With the presence of ammonium formate both in the mobile phase and in the make-up solvent, the *m/z* value of the ion for TAGs was $[M+NH_4]^+$ (positive ESI; cone voltage equal to 35 V). The programmed cone voltage fragmentation was applied to get fragment ions assigned to DAG and MAG, respectively, at 50 and 90 V, allowing the TAG identification [102]. The effect of the postcolumn addition of lithium iodine (in methanol) on the MS signal response of acetogenins was recently studied. This addition allowed for improved structural information compared to the classical sodium adducts, but also reduced the response intensity, t the sensitivity of the method [40].

With an APCI source, both molecular ($[LLL + H]^+$) and fragment ions ($[LL + H]^+$ and $[(L + H) - H_2O]^+$) were produced from TAGs [94]. An extensive study of the make-up solvent addition with APCI/Q-ToF (quadrupole time of flight) [29], including the location of the make-up versus the BPR, the make-up flow rate, and the make-up solvent nature (isopropylalcohol, MeOH, MeOH + 0.1% formic acid), showed the positive effect of the make-up flow rate up to 0.5 mL/min for the TAG and DAG and the higher response observed with MeOH–formic acid. In that case, the pre- or post-BPR location was ineffective on the response.

Three different ionization sources, ESI, APCI, and APPI, were compared for the analysis of tocopherols. The APPI source displayed the highest response when using ethanol as the make-up solvent without any dopant (0.2 mL/min) [11].

Applied to real samples, strong suppression effects were reported for the analysis of lipopolysaccharides from the cell wall of Gram-negative bacteria, which is able to cause severe inflammation [100].

The sensitivity of the SFC-MS coupling is often studied in terms of limit of detection and limit of quantification (LOQ). A comparison for phospholipids showed that the LOQ was about 20 ng/g for the polyene phosphatidylcholine with SFC-MS/MS and

60 ng/g with HPLC-ELSD. However, the column and particle size dimensions and the analyses duration were considerably different and could explain in part the difference in LOQ values [103].

The ELSD is another open-cell detector used to couple with SFC. This coupling does not require any special interface, and the presence of carbon dioxide in the mobile phase favors the intensity of the chromatographic peaks in SFC versus HPLC [105].

ELSD was largely used for the analysis of monosaccharides and polyols with varied polar SPs (e.g., cyano, amino, diol, and silica) [106, 107]. This detector allowed the use of cosolvent gradient and required to elute mono-, di-, and trisaccharides without any baseline drift [87].

The reported advantages of the SFC-ELSD for the analysis of ginkgolide A, B, C, J and bilobalide (terpene trilactones from *Ginkgo biloba* leaves) were: a higher sensitivity in comparison to UV; better baseline stability in comparison to HPLC-refractive index; and no need of derivatization in comparison with GC with flame ionization detection [108]. ELSD was also used for the analysis of ginsenosides, which are polar triterpenoid saponins from *Panax quinquefolius*. The ELSD was located before the BPR with a split, allowing the collection of fractions after the BPR for further structural identification with NMR [109]. The introduction of varied additives (TFA, di-IPA, and aqueous ammonium acetate) in the methanol cosolvent resulted in decreased signal-to-noise ratio of the ginsenosides. Finally, only 0.05% of TFA was added to the modifier [109].

A split between ELSD and BPR was also used, with a 1:3 ratio for the analysis of phospholipids. This analysis is very challenging with SFC because the polar functional group of these compounds is strongly retained by polar SPs despite a high content of methanol in carbon dioxide. An octyl-bonded phase, along with a strong elution gradient and 0.1% TFA in the cosolvent to neutralize the charged phosphate group, was used to solve this issue [75]. Recently, a 0.8:2 split was used for the study of polar lipids, cerebrosides, monogalactosyl diglyceride, and digalactosyl diglycerides from deoiled soybean lecithin [110]. Some of these compounds (monogalactosyl diglyceride and digalactosyl diglycerides) are also found in wheat, and a profiling of glucolipids classes with SFC-ELSD was achieved with a methanol gradient [111]. A strong enhancement of the SFC-ELSD lipid response was reported when adding an equimolar mixture of trimethylamine and formic acid (1% in dichloromethane) at 0.1 mL/min after the column and before the BPR. The postcolumn addition was required because of the nature of the SP (silica), for which formic acid was supposedly unsuitable [112]. In another case, working with a porous graphitic carbon SP, these additives could be directly added into the mobile phase cosolvent to produce a strong response improvement [113].

Finally, a satisfactory separation and detection of hydroxyl and acidic triterpenoids (ursane, lupane, and oleane) from apple peel was achieved with a CO_2–methanol 97:3 (v/v) mixture on phenyl-oxypropyl-bonded SP (Synergi Polar RP). With such low solvent proportion, the ELSD response is significantly improved [64]. Figure 7.5 shows the separation of eight standard compounds, and extracts of four apple varieties. The three acidic triterpenoids (oleanolic, ursolic, and betulinic acids) were present in all the extracts, with small amounts of β-amyrin and lupeol. Additional peaks were also detected, but due to the presence of many positional isomers having the same molecular weight, for instance $m/z = 472$ or 488, no definitive attribution could be done by MS from eight other acidic triterpenoids. Another study was carried out on crude extracts of the leaves of *Rosa sericea*, which also contain numerous aglycone triterpenoids and triterpene glycosides [114]. With a linear gradient (1–35%) of methanol containing 0.1% of TFA, the LOQ for acidic triterpenoids and β-sitosterol ranged from 1.35 to4.55 µg/mL.

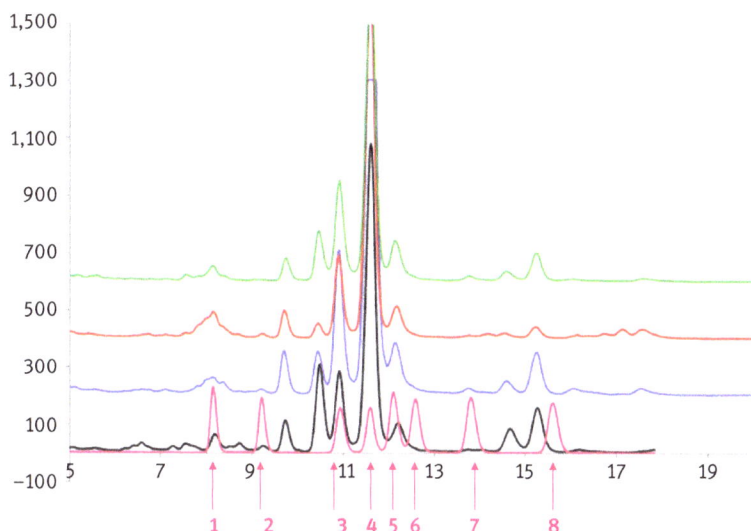

Figure 7.5: Separation of standards of triterpenoids (bottom chromatogram) and four samples of apple peel extracts (Royal Gala, Golden, Granny Smith, Pink Lady). Synergi Polar RP (250 × 4.6 mm, 4 µm), carbon dioxide–methanol 97:3 (v/v), 20 °C, 12 MPa, ELSD. Standards : (1) β-amyrin, (2) lupeol, (3) oleanolic acid, (4) ursolic acid, (5) betulinic acid, (6) erythrodiol, (7) uvaol, and (8) betulin [64].

For several reasons, ELSD is also very useful for the study of cosmetic products, which are mostly based on oil–water emulsions. First, many ingredients included in cosmetic products have no chromophoric groups; thus, ELSD avoids the need to derivatize for UV or FID detection [115]. Secondly, although the performance of ELSD is rather poor at low concentrations in comparison to UV or MS, the linear range reaches quite high concentrations, which is very useful to quantify ingredients that may be present in large amounts in a cosmetic formulation. Thirdly, when using RPLC, the presence of lipids in the oil–water emulsions requires the use of a large amount of organic solvent at the end of the analysis to flush the column, whereas with the supercritical fluids, the solubility of these compounds is favored.

Two studies have reported on the use of SFC-ELSD to analyze cosmetic waxes (candellila or carnauba) [116], white beeswax, and hydrogenated jojoba oil [117]. Composition anomalies could be observed related to the overabundance of certain compounds.

Recently a paper showed the large application range of SFC-ELSD for the analyses of ingredients in body creams or eye-liner [118]. One of the three examples was related to the quantitation of glyceryl caprylate, an emollient produced from vegetable oils that does not adsorb UV light. Figure 7.6 shows the comparative UV (210 nm) and ELSD responses for the samples, obtained with an HSS C18 SB 5 μm column and an isocratic elution (CO_2–methanol 95:5 v/v). The principal ingredient was not detected with UV, whereas additional matrix peaks were separated in less than 4 min.

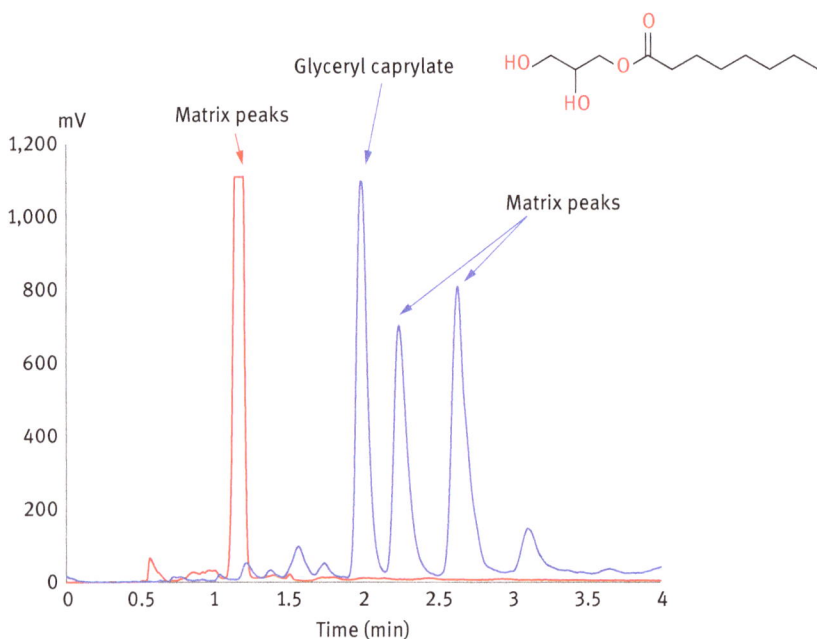

Figure 7.6: Analysis of an eye serum cosmetic cream with UV 210 nm (red trace) or ELSD (blue trace) to quantify glyceryl caprylate. HSS C18 SB (250 × 4.6 mm, 5 μm), carbon dioxide–methanol 95:5 (v/v)

7.5 Semipreparative SFC of natural products and off-line two-dimensional separation

Several examples of fractionation of plant extracts were achieved with semiprepa-rative SFC prior to NMR studies or with the aim to provide standard molecules. The volume of solvent collected for each fraction is moderate because only the modifier is collected, while carbon dioxide depressurizes after the BPR and is evacuated as a gas. However, a make-up pump is sometimes necessary after the BPR to increase the flow of liquid, favoring the solubility of compounds in the transfer capillary to the fraction collector [41, 73].

Sometimes, analytical columns are simply used with large injection volume, in the case of simple separations with good resolution. For instance, carnosic acid and carnosol from rosemary extracts were purified in 40 batch runs, with 100 μL injected per run [41]. In another example, falcarinol and falcarindiol found in carrots, celery, and parsley were purified in 85 runs with 70 μL injected per run [73].

Here again, coupled analytical columns may be used. For instance, 40 cm of Chiralpak IC was used for the separation of the 25 (*R/S*)-spirostanol saponin diastereoisomers from *Trigonella foenum-graecum* seeds (five stacked injections of 70 μL each) [119]. Another example was given by the use of two C4-bonded columns (protein C4) for the separation of seven compounds from the extracts of Kava lactone roots (*V*inj = 50 μL) [120].

Semipreparative or preparative columns (with larger internal diameter) were also used (Table 7.2) [42, 70, 79, 121–125]. Scale-up methods are usually achieved from previously developed analytical methods (increasing both flow rate and injected volume). A very subtle recycle preparative method was described for the collection of tocopherols from wheat germ oil [126]. In this system, after the first column, a switching valve was connected to the BPR and was allowed to send the eluent either to the waste or to a second column to retain compounds having weak retention, which could be sent again by recycling into the first column (Figure 7.7). With two silica SPs, the first portion (TAGs) of the oil (not retained by the silica) was sent to the waste/collection, the second portion containing α- and β-tocopherols was retained in the second column by switching the valve, and the third fraction from the first column was also sent to the waste/collection. Then, the second fraction that was sent to the second column was sent again into the first column, allowing a better resolution of the two tocopherol isomers. The number of recycle runs depends on the separation quality required before collecting the fractions. This fractionation was achieved prior to further analyses of the collected fractions, either by GC or RPLC.

Off-line two-dimensional SFC × LC or LC × SFC analyses were investigated using either preparative or analytical column in both dimensions. The off-line mode avoids the need for a specific interface to transfer a liquid (or supercritical fluid) to a supercritical fluid (or liquid) because of the difficulty of mixing two mobile phases with different compressibility [127]. The separation of blackberry sage fragrant oil was studied by RPLC (C18) × SFC (amino) off-line system with a gradient in the two dimensions (water–acetonitrile for RPLC and CO_2–acetonitrile for SFC). In the first dimension, operated at 0.5 mL/min (*V*inj = 80 μL), 56 fractions were collected prior to their SFC analysis. The total duration of the first dimension was 18 min. A 50 μL aliquot of each fraction was analyzed by SFC, with a flow rate of 3 mL/min. The theoretical peak capacity was equal to 4,232, whereas the practical one was 2,405, with a total analysis duration of 280 min.

Table 7.2: Sample applications of preparative SFC.

Analytes, sample	SP	Column dimensions	Flow rate	Injected volume	Run number	Total duration	Fraction number	Purified amount	Ref.
Curcuminoids	Silica	250 × 19 mm	80 mL/min	100 µL	22	5.5 hr	3	32.4 g	[80]
Davanone	2-EP	250 × 10 mm	4 mL/min	40 µL			3		[121]
Phenolic styrax	2-EP	250 × 15 mm	15 mL/min +5 mL/min make-up	250 µL			6		[42]
Osthole/Impe-ratorin	NH2	250 × 10 mm	20 mL/min	200 µL	20		2	11 mg	[122]
Lignan dia.	Polysaccharide	250 × 20 mm	60 g/min	1,500 µL	14	37 min SI*	2	113.1 mg	[70]
Lemon peel oil	Silica	50 × 7.2 mm	2.2 g/min	500 µL			4		[123]
Crude palm oil	C18	100 × 10 mm		500 µL			4		[124]
Lignans	Xamide	250 × 20 mm	50 mL/min	2,000 µL			10	15.5 g	[125]

SI*: stacked injections.

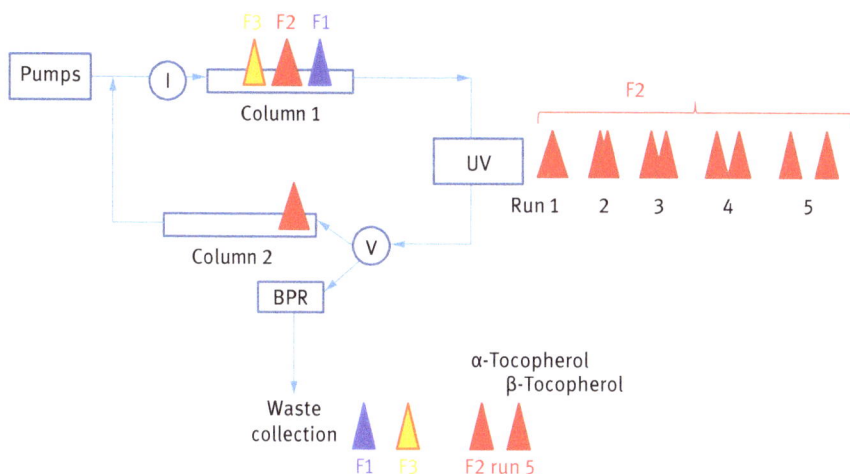

Figure 7.7: Design of an original preparative recycling system to collect critical pairs of peaks with high purity [126].

The separation of amine alkaloids from *Piper longum* was studied by off-line SFC × UHPLC [128]. Thirty-two fractions were collected (each of them during 30 sec) from the first SFC dimension on an amide column and analyzed by RPLC with an HSS C18 T3 phase. The analysis duration in the first dimension was 20 min, and 30 min for each fraction in the second dimension, leading to a total analysis time of 620 min. The fraction containing pigments (carotenes, xanthophylls, and chlorophylls) in sweet bell peppers (*Capsicum annuum* L.) was also analyzed by SFC × RPLC with an HSS C18 SB phase in the first dimension and a YMC C30 phase in the second dimension [129].

An off-line HILIC-SFC-RPLC approach was used to analyze lignans (with anti-inflammatory, antitumor, and antidiabetic activity) from *Fructus arctii* [125]. A HILIC system using a cysteine-bonded silica SP was first used to obtain three fractions. The fractions were further divided into 10 fractions with an SFC analysis in 20 min on a semipreparative Xamide column (250 × 20 mm). Finally, each fraction was analyzed with RPLC on a C18 phase in 30 min. This three-step approach supplied 12 purified compounds for MS and NMR characterization.

Off-line SFC–NARP-LC (nonaqueous RPLC) was applied to the analysis of TAGs from fish oil [130], using two sulfonic silica-based strong cation-exchange columns coated with silver in the first SFC dimension (silver ion or SI mode), and three Zorbax SB C18 columns in NARP-LC in the second dimension. From the 150 min analysis in NARP-LC of 18 fractions collected in SI-SFC between 40 and 150 min, 324 TAGs could be identified.

Only a few on-line two-dimensional methods were reported. The first paper describing SFC × SFC [131] used two C18 phases in each dimension, neat CO_2, and two different temperature values to separate components of perrila oil: 10 °C in the first dimension and 50 °C in the second one. A metal capillary column (0.25 mm internal diameter, 100% methylsilicone, 0.25 μm film thickness, and 100 cm length) was used as the trapping tube between the two dimensions, in which the compounds were stacked during the complete evaporation of neat CO_2. Because only one trap was used, the first dimension was operated in the stop-flow mode. This system was also applied to the analysis of free fatty acids of fish oil [132], with a bare silica phase in the first dimension and a C18 phase in the second one.

An interface using trapping columns was used for the SFC × RPLC analysis of lemon oil extract [133]. A cyanopropyl-bonded silica phase (250 × 2 mm) was used in the first dimension and a Zorbax SB C18 phase (50 × 4.6 mm) in the second dimension. The interface allowed introducing water (1.5 mL/min) to the supercritical phase (0.5 mL/min) eluting from the first dimension, with the aim to focus the analytes onto a C18 phase located in a small (7.2 × 4.6 mm) trapping column set in a loop before the second dimension, operated at high flow rate (4.6 mL/min). This small packed column, located after the BPR, allowed trapping the compounds entering the loop in a plug of expanded CO_2, that is, with a high linear speed. The analysis duration in the SFC first dimension was 65 min and was 0.7 min in the second RPLC dimension. By comparing to an NPLC × RPLC system, this SFC × RPLC system favored the performance of the second dimension because no hazardous solvent (hexane) was needed to be introduced in the second dimension. The same interface was also used for TAG analysis of fish oil; this time coupling SI-SFC to NARP-LC [130]. In that case, compared to the off-line approach described previously, a monolithic Chromolith RP-18 column was used to reduce the analysis duration down to 2 min in the second dimension.

However, when comparing comprehensive SFC × SFC and high-efficiency SFC (with coupled columns), for instance, for the analysis of furocoumarins in lemon [62, 134] or TAG of edible oils [27, 133], the increase in separation performance expected from the SFC × SFC approach was not significant enough, despite a greater instrument complexity.

7.6 Conclusion

The use of CO_2-based fluids as mobile phase presents numerous advantages for many application fields, including that of natural products. SFC offers varied possibilities to solve modern issues. Recently, the simultaneous analysis of water- and fat-soluble vitamins was achieved on an HSS C18 SB column with a gradient elution of 5–100% cosolvent (methanol–water containing 0.2% ammonium formate), allowing the separation of 17 vitamins in 4 min [135]. SFC is being increasingly utilized for chemical profiling of lipids [136], bioactive compounds [137], and plant extracts [138].

Obviously, the easy coupling to MS is a powerful tool to identify target compounds in complex extracts. However, numerous variables should be carefully adjusted, for instance, using experimental design, from the make-up composition and flow rate to the ionization source parameters in the goal to favor ionization [104]. In addition, conclusions on the optimal parameters depend on both the analytes and ionization source (ESI, APCI, or APPI). Moreover, when working with crude plant extracts, strong matrix effects (signal suppression) may render the calibration with pure standards unreliable and may yield negative effects on limits of quantitation.

Evaporative light-scattering detection is also useful for the analysis of cosmetic product containing high amounts of ingredients that do not absorb UV light. Its sensitivity is favored by the use of pressurized carbon dioxide and low proportions of cosolvent.

The variety of SPs and the understanding of their behavior through the interaction map developed on the basis of the LSER model offer numerous possibilities to achieve separations of mixtures of natural compounds. Nevertheless, it appears that bare silica, HSS C18 SB, and 2-EP phases were most often selected since they all provided high selectivity and symmetrical peaks for numerous mixtures containing neutral, basic, and acidic natural compounds. In the case of late elution of nonsymmetrical peaks, additives can be added to the cosolvent to improve both the chromatographic separation and the detection response, particularly with MS detection.

Full identification of analytes requires NMR experiments, that is, the ability to collect a sufficient quantity of purified compound. Fortunately, SFC allows easy scale-up from analytical to semipreparative separation.

Besides, supercritical fluids can also be used to extract natural products, and selective extraction can be achieved due to the variability of modifier chemical nature and amount that can be added to carbon dioxide [28]. Impregnation of bioactive compounds onto solid particles for cosmetic formulation is also being investigated [139], and the nature of supercritical fluids favors on-line extraction–impregnation processes.

Super flexible combination of operating modules should be developed in the future to achieve varied on-line operations: dilution–separation–quantification for cosmetics, extraction–separation–identification for bioactive analytes from plant extracts, and selective extraction–impregnation for new formulations.

References

[1] Lesellier, E., Overview of the retention in subcritical fluid chromatography with varied polarity stationary phases. *J. Sep. Sci.* 2008, 31, 1238–1251.
[2] Lesellier, E., Retention mechanisms in super/subcritical fluid chromatography on packed columns. *J. Chromatogr. A* 2009, 1216, 1881–1890.
[3] West, C., Lesellier, E., Effects of mobile phase composition on retention and selectivity in achiral supercritical fluid chromatography. *J. Chromatogr. A* 2013, 1302, 152–162.
[4] Lesellier, E., West, C., The many faces of packed column supercritical fluid chromatography – A critical review. *J. Chromatogr. A* 2015, 1382, 2–46.
[5] Nováková, L., Grand-Guillaume Perrenoud, A., Francois, I., West, C., Lesellier, E., Guillarme, D., Modern analytical supercritical fluid chromatography using columns packed with sub-2 μm particles: A tutorial. *Anal. Chim. Acta* 2014, 824, 18–35.

[6] West, C., Khater, S., Lesellier, E., Characterization and use of hydrophilic interaction liquid chromatography type stationary phases in supercritical fluid chromatography. *J. Chromatogr. A* 2012, 1250, 182–195.

[7] Pourmortazavi, S. M., Hajimirsadeghi, S. S., Supercritical fluid extraction in plant essential and volatile oil analysis. *J. Chromatogr. A* 2007, 1163, 2–24.

[8] de Melo, M. M. R., Silvestre, A. J. D., Silva, C. M., Supercritical fluid extraction of vegetable matrices: Applications, trends and future perspectives of a convincing green technology. *J. Supercrit. Fluids* 2014, 92, 115–176.

[9] Yarita, T., Nomura, A., Abe, K., Takeshita, Y., Supercritical fluid chromatographic determination of tocopherols on an ODS-silica gel column. *J. Chromatogr. A* 1994, 679, 329–334.

[10] Han, N. M., May, C. Y., Ngan, M. A., Hock, C. C., Hashim, M. A., Isolation of palm tocols using supercritical fluid chromatography. *J. Chromatogr. Sci.* 2004, 42, 536–539.

[11] Méjean, M., Brunelle, A., Touboul, D., Quantification of tocopherols and tocotrienols in soybean oil by supercritical-fluid chromatography coupled to high-resolution mass spectrometry. *Anal. Bioanal. Chem.* 2015, 407, 5133–5142.

[12] Ibáñez, E., Palacios, J., Señoráns, F. J., Santa-María, G., Tabera, J., Reglero, G., Isolation and separation of tocopherols from olive by-products with supercritical fluids. *J. Am. Oil Chem. Soc.* 2000, 77, 187–190.

[13] Esteban, R., Moran, J. F., Becerril, J. M., García-Plazaola, J. I., Versatility of carotenoids: An integrated view on diversity, evolution, functional roles and environmental interactions. *Environ. Exp. Bot.* 2015, 119, 63–75.

[14] Aubert, M.-C., Lee, C. R., Krstuloviác, A. M., Lesellier, E., Péchard, M.-R., Tchapla, A., Separation of trans/cis- α- and β-carotenes by supercritical fluid chromatography: I. Effects of temperature, pressure and organic modifiers on the retention of carotenes. *J. Chromatogr. A* 1991, 557, 47–58.

[15] Lesellier, E., Tchapla, A., Péchard, M.-R., Lee, C. R., Krstulović, A. M., Separation of trans/cis α- and β-carotenes by supercritical fluid chromatography: II. Effect of the type of octadecyl-bonded stationary phase on retention and selectivity of carotenes. *J. Chromatogr. A* 1991, 557, 59–67.

[16] Matsubara, A., Bamba, T., Ishida, H., Fukusaki, E., Hirata, K., Highly sensitive and accurate profiling of carotenoids by supercritical fluid chromatography coupled with mass spectrometry. *J. Sep. Sci.* 2009, 32, 1459–1464.

[17] Lesellier, E., Tchapla, A., A simple subcritical chromatographic test for an extended ODS high performance liquid chromatography column classification. *J. Chromatogr. A* 2005, 1100, 45–59.

[18] Lesellier, E., West, C., Tchapla, A., Classification of special octadecyl-bonded phases by the carotenoid test. *J. Chromatogr. A* 2006, 1111, 62–70.

[19] Buskov, S., Sørensen, H., Sørensen, S., Separation of chlorophylls and their degradation products using packed column Supercritical Fluid Chromatography (SFC). *J. High Resolut. Chromatogr.* 1999, 22, 339–342.

[20] Abrahamsson, V., Rodriguez-Meizoso, I., Turner, C., Determination of carotenoids in microalgae using supercritical fluid extraction and chromatography. *J. Chromatogr. A* 2012, 1250, 63–68.

[21] Agrawal, H., Kaul, N., Paradkar, A. R., Mahadik, K. R., Standardization of crude extract of neem seed kernels (Azadirachta indica A. Juss) and commercial neem based formulations using HPTLC and extended length packed-columns SFC Method. *Chromatographia* 2005, 62, 183–195.

[22] Bartolomé Ortega, A., Škerget, M., Knez, Ž., Separation of xanthohumol from hop extracts by supercritical fluid chromatography. *Chem. Eng. Res. Des.* 2016, 109, 335–345.

[23] Bamba, T., Fukusaki, E., Kajiyama, S., Ute, K., Kitayama, T., Kobayashi, A., High-resolution analysis of polyprenols by supercritical fluid chromatography. *J. Chromatogr. A* 2001, 911, 113–117.

[24] Lesellier, E., Bleton, J., Tchapla, A., Use of relationships between retention behaviors and chemical structures in subcritical fluid chromatography with CO_2/modifier mixtures for the identification of triglycerides. *Anal. Chem.* 2000, 72, 2573–2580.

[25] Lesellier, E., Tchapla, A., Separation of vegetable oil triglycerides by subcritical fluid chromatography with octadecyl packed columns and CO_2/modifier mobile phases. *Chromatographia* 2000, 51, 688–694.

[26] Lesellier, E., Efficiency in supercritical fluid chromatography with different superficially porous and fully porous particles ODS bonded phases. *J. Chromatogr. A* 2012, 1228, 89–98.

[27] Lesellier, E., Latos, A., de Oliveira, A. L., Ultra high efficiency/low pressure supercritical fluid chromatography with superficially porous particles for triglyceride separation. *J. Chromatogr. A* 2014, 1327, 141–148.

[28] Duval, J., Destandau, E., Pecher, V., Poujol, M., Tranchant, J.-F., Lesellier, E., Selective enrichment in bioactive compound from Kniphofia uvaria by super/subcritical fluid extraction and centrifugal partition chromatography. *J. Chromatogr. A* 2016, 1447, 26–38.

[29] Duval, J., Colas, C., Pecher, V., Poujol, M., Tranchant, J.-F., Lesellier, E., Hyphenation of ultra high performance supercritical fluid chromatography with atmospheric pressure chemical ionisation high resolution mass spectrometry: Part 1. Study of the coupling parameters for the analysis of natural non-polar compounds. *J. Chromatogr. A* 2017, 1509, 132–140.

[30] Huang, Y., Feng, Y., Tang, G., Li, M., Zhang, T., Fillet, M., Crommen, J., Jiang, Z., Development and validation of a fast SFC method for the analysis of flavonoids in plant extracts. *J. Pharm. Biomed. Anal.* 2017, 140, 384–391.

[31] Berger, T. A., Berger, B. K., Separation of natural food pigments in saponified and un-saponified paprika oleoresin by Ultra High Performance Supercritical Fluid Chromatography (UHPSFC). *Chromatographia* 2013, 76, 591–601.

[32] Huang, Y., Zhang, T., Zhou, H., Feng, Y., Fan, C., Chen, W., Crommen, J., Jiang, Z., Fast separation of triterpenoid saponins using supercritical fluid chromatography coupled with single quadrupole mass spectrometry. *J. Pharm. Biomed. Anal.* 2016, 121, 22–29.

[33] Jumaah, F., Plaza, M., Abrahamsson, V., Turner, C., Sandahl, M., A fast and sensitive method for the separation of carotenoids using ultra-high performance supercritical fluid chromatography-mass spectrometry. *Anal. Bioanal. Chem.* 2016, 408, 5883–5894.

[34] Yang, J., Zhu, L., Zhao, Y., Xu, Y., Sun, Q., Liu, S., Liu, C., Ma, B., Separation of furostanol saponins by supercritical fluid chromatography. *J. Pharm. Biomed. Anal.* 2017, 145, 71–78.

[35] Kamangerpour, A., Ashraf-Khorassani, M., Taylor, L. T., McNair, H. M., Chorida, L., Supercritical fluid chromatography of polyphenolic compounds in grape seed extract. *Chromatographia* 2002, 55, 417–421.

[36] Murauer, A., Ganzera, M., Quantitative determination of lactones in piper methysticum (Kava-Kava) by supercritical fluid chromatography. *Planta Med.* 2017, 83, 1053–1057.

[37] Ganzera, M., Supercritical fluid chromatography for the separation of isoflavones. *J. Pharm. Biomed. Anal.* 2015, 107, 364–369.

[38] Tu, A., Du, Z., Qu, S., Rapid profiling of triacylglycerols for identifying authenticity of edible oils using supercritical fluid chromatography-quadruple time-of-flight mass spectrometry combined with chemometric tools. *Anal Methods* 2016, 8, 4226–4238.

[39] Ovchinnikov, D. V., Kosyakov, D. S., Ul'yanovskii, N. V., Bogolitsyn, K. G., Falev, D. I., Pokrovskiy, O. I., Determination of natural aromatic acids using supercritical fluid chromatography. *Russ. J. Phys. Chem. B* 2016, 10, 1062–1071.

[40] Laboureur, L., Bonneau, N., Champy, P., Brunelle, A., Touboul, D., Structural characterisation of acetogenins from *Annona muricata* by supercritical fluid chromatography coupled to

high-resolution tandem mass spectrometry: Structural characterization of acetogenins by SFC-HRMS/MS. *Phytochem. Anal.* 2017, 28, 512–520

[41] Vicente, G., García-Risco, M. R., Fornari, T., Reglero, G., Isolation of carsonic acid from rosemary extracts using semi-preparative supercritical fluid chromatography. *J. Chromatogr. A* 2013, 1286, 208–215.

[42] Scheuba, J., Wronski, V.-K., Rollinger, J., Grienke, U., Fast and green – CO_2 based extraction, isolation, and quantification of phenolic styrax constituents. *Planta Med.* 2017, 83, 1068–1075.

[43] Wu, W., Zhang, Y., Wu, H., Zhou, W., Cheng, Y., Li, H., Zhang, C., Li, L., Huang, Y., Zhang, F., Simple, rapid, and environmentally friendly method for the separation of isoflavones using ultra-high performance supercritical fluid chromatography. *J. Sep. Sci.* 2017, 40, 2827–2837.

[44] Wang, B., Liu, X., Zhou, W., Hong, Y., Feng, S., Fast separation of flavonoids by supercritical fluid chromatography using a column packed with a sub-2 µm particle stationary phase. *J. Sep. Sci.* 2017, 40, 1410–1420.

[45] Kim, H. S., Chun, J. M., Kwon, B.-I., Lee, A.-R., Kim, H. K., Lee, A. Y., Development and validation of an ultra-performance convergence chromatography method for the quality control of Angelica gigas Nakai. *J. Sep. Sci.* 2016, 39, 4035–4041.

[46] Kim, H. S., Moon, B. C., Choi, G., Kim, W. J., Lee, A. Y., Ultra-performance convergence chromatography for the quantitative determination of bioactive compounds in *Aralia continentalis* Kitagawa as quality control markers. *J. Sep. Sci.* 2017, 40, 2071–2079.

[47] Winderl, B., Schwaiger, S., Ganzera, M., Fast and improved separation of major coumarins in *Ammi visnaga* (L.) Lam. by supercritical fluid chromatography: Other Techniques. *J. Sep. Sci.* 2016, 39, 4042–4048.

[48] Gong, X., Qi, N., Wang, X., Lin, L., Li, J., Ultra-performance convergence chromatography (UPC2) method for the analysis of biogenic amines in fermented foods. *Food Chem.* 2014, 162, 172–175.

[49] Qu, S., Du, Z., Zhang, Y., Direct detection of free fatty acids in edible oils using supercritical fluid chromatography coupled with mass spectrometry. *Food Chem.* 2015, 170, 463–469.

[50] Aichner, D., Ganzera, M., Analysis of anthraquinones in rhubarb (Rheum palmatum and Rheum officinale) by supercritical fluid chromatography. *Talanta* 2015, 144, 1239–1244.

[51] Giuffrida, D., Zoccali, M., Giofrè, S. V., Dugo, P., Mondello, L., Apocarotenoids determination in Capsicum chinense Jacq. cv. Habanero, by supercritical fluid chromatography-triple-quadrupole/mass spectrometry. *Food Chem.* 2017, 231, 316–323.

[52] Heaton, D. M., Bartle, K. D., Rayner, C. M., Clifford, A. A., Application of supercritical fluid extraction and supercritical fluid chromatography to the production of taxanes as anti-cancer drugs. *J. High Resolut. Chromatogr.* 1993, 16, 666–670.

[53] West, C., Lesellier, E., Characterization of stationary phases in subcritical fluid chromatography by the solvation parameter model. *J. Chromatogr. A* 2006, 1110, 181–190.

[54] West, C., Lesellier, E., Characterisation of stationary phases in subcritical fluid chromatography by the solvation parameter model. *J. Chromatogr. A* 2006, 1110, 191–199.

[55] West, C., Lesellier, E., Characterisation of stationary phases in subcritical fluid chromatography with the solvation parameter model. *J. Chromatogr. A* 2006, 1110, 200–213.

[56] West, C., Lesellier, E., Characterisation of stationary phases in subcritical fluid chromatography with the solvation parameter model IV. *J. Chromatogr. A* 2006, 1115, 233–245.

[57] West, C., Lesellier, E., Characterisation of stationary phases in supercritical fluid chromatography with the solvation parameter model: V. Elaboration of a reduced set of test solutes for rapid evaluation. *J. Chromatogr. A* 2007, 1169, 205–219.

[58] West, C., Lesellier, E., Orthogonal screening system of columns for supercritical fluid chromatography. *J. Chromatogr. A* 2008, 1203, 105–113.

[59] West, C., Lesellier, E., A unified classification of stationary phases for packed column supercritical fluid chromatography. *J. Chromatogr. A* 2008, 1191, 21–39.

[60] West, C., Lemasson, E., Khater, S., Lesellier, E., An attempt to estimate ionic interactions with phenyl and pentafluorophenyl stationary phases in supercritical fluid chromatography. *J. Chromatogr. A* 2015, 1412, 126–138.

[61] West, C., Lemasson, E., Bertin, S., Hennig, P., Lesellier, E., An improved classification of stationary phases for ultra-high performance supercritical fluid chromatography. *J. Chromatogr. A* 2016, 1440, 212–228.

[62] Desmortreux, C., Rothaupt, M., West, C., Lesellier, E., Improved separation of furocoumarins of essential oils by supercritical fluid chromatography. *J. Chromatogr. A* 2009, 1216, 7088–7095.

[63] Pfeifer, I., Murauer, A., Ganzera, M., Determination of coumarins in the roots of Angelica dahurica by supercritical fluid chromatography. *J. Pharm. Biomed. Anal.* 2016, 129, 246–251.

[64] Lesellier, E., Destandau, E., Grigoras, C., Fougère, L., Elfakir, C., Fast separation of triterpenoids by supercritical fluid chromatography/evaporative light scattering detector. *J. Chromatogr. A* 2012, 1268, 157–165.

[65] West, C., Lesellier, E., Separation of substituted aromatic isomers with porous graphitic carbon in subcritical fluid chromatography. *J. Chromatogr. A* 2005, 1099, 175–184.

[66] Berger, T. A., Berger, B. K., Chromatographic resolution of 7 of 8 stereoisomers of vitamin K1 on an amylose stationary phase using supercritical fluid chromatography. *Chromatographia* 2013, 76, 549–552.

[67] Xie, J., Cheng, J., Han, H., Sun, B., Yanik, G. W., Resolution of racemic γ-lactone flavors on Chiralpak AD by packed column supercritical fluid chromatography. *Food Chem.* 2011, 124, 1107–1112.

[68] Schaffrath, M., Weidmann, V., Maison, W., Enantioselective high performance liquid chromatography and supercritical fluid chromatography separation of spirocyclic terpenoid flavor compounds. *J. Chromatogr. A* 2014, 1363, 270–277.

[69] Krief, A., Dunkle, M., Bahar, M., Bultinck, P., Herrebout, W., Sandra, P., Elucidation of the absolute configuration of rhizopine by chiral supercritical fluid chromatography and vibrational circular dichroism: Other techniques. *J. Sep. Sci.* 2015, 38, 2545–2550.

[70] Xin, H., Dai, Z., Cai, J., Ke, Y., Shi, H., Fu, Q., Jin, Y., Liang, X., Rapid purification of diastereoisomers from Piper kadsura using supercritical fluid chromatography with chiral stationary phases. *J. Chromatogr. A* 2017, 1509, 141–146.

[71] Qiao, X., An, R., Huang, Y., Ji, S., Li, L., Tzeng, Y., Guo, D., Ye, M., Separation of 25R/S-ergostane triterpenoids in the medicinal mushroom Antrodia camphorata using analytical supercritical-fluid chromatography. *J. Chromatogr. A* 2014, 1358, 252–260.

[72] Lee, J. W., Nagai, T., Gotoh, N., Fukusaki, E., Bamba, T., Profiling of regioisomeric triacylglycerols in edible oils by supercritical fluid chromatography/tandem mass spectrometry. *J. Chromatogr. B* 2014, 966, 193–199.

[73] Bijttebier, S., D'Hondt, E., Noten, B., Hermans, N., Apers, S., Exarchou, V., Voorspoels, S., Automated analytical standard production with supercritical fluid chromatography for the quantification of bioactive C17-polyacetylenes: A case study on food processing waste. *Food Chem.* 2014, 165, 371–378.

[74] Berger, T. A., Berger, B. K., Two minute separation of the cis- and trans-isomers of vitamin K1 without heptane, chlorinated solvents, or acetonitrile. *Chromatographia* 2013, 76, 109–115.

[75] Eckard, P. R., Taylor, L. T., Slack, G. C., Method development for the separation of phospholipids by subcritical fluid chromatography. *J. Chromatogr. A* 1998, 826, 241–247.

[76] Buskov, S., Hasselstrøm, J., Olsen, C. E., Sørensen, H., Sørensen, J. C., Sørensen, S., Supercritical fluid chromatography as a method of analysis for the determination of 4-hydroxybenzylglucosinolate degradation products. *J. Biochem. Biophys. Methods* 2000, 43, 157–174.

[77] Buskov, S., Hansen, L. B., Olsen, C. E., Sørensen, J. C., Sørensen, H., Sørensen, S., Determination of ascorbigens in autolysates of various brassica species using supercritical fluid chromatography. *J. Agric. Food Chem.* 2000, 48, 2693–2701.

[78] Buskov, S., Olsen, C. E., Sørensen, H., Sørensen, S., Supercritical fluid chromatography as basis for identification and quantitative determination of indol-3-ylmethyl oligomers and ascorbigens. *J. Biochem. Biophys. Methods* 2000, 43, 175–195.

[79] Song, W., Qiao, X., Liang, W., Ji, S., Yang, L., Wang, Y., Xu, Y., Yang, Y., Guo, D., Ye, M., Efficient separation of curcumin, demethoxycurcumin, and bisdemethoxycurcumin from turmeric using supercritical fluid chromatography: From analytical to preparative scale: Sample Preparation. *J. Sep. Sci.* 2015, 38, 3450–3453.

[80] Ashraf-Khorassani, M., Taylor, L. T., Marr, J. G. D., Analysis of the sulfomycin component of alexomycin in animal feed by enhanced solvent extraction and supercritical fluid chromatography. *J. Biochem. Biophys. Methods* 2000, 43, 147–156.

[81] Janicot, J. L., Caude, M., Rosset, R., Separation of opium alkaloids by carbon dioxide sub- and supercritical fluid chromatography with packed columns : Application to the quantitative analysis of poppy straw extracts. *J. Chromatogr. A* 1988, 437, 351–364.

[82] Dost, K., Davidson, G., Development of a packed-column supercritical fluid chromatography/atmospheric pressure chemical-ionisation mass spectrometric technique for the analysis of atropine. *J. Biochem. Biophys. Methods* 2000, 43, 125–134.

[83] Huang, Y., Zhang, T., Zhao, Y., Zhou, H., Tang, G., Fillet, M., Crommen, J., Jiang, Z., Simultaneous analysis of nucleobases, nucleosides and ginsenosides in ginseng extracts using supercritical fluid chromatography coupled with single quadrupole mass spectrometry. *J. Pharm. Biomed. Anal.* 2017, 144, 213–219.

[84] Bicchi, C., Balbo, C., Rubiolo, P., Packed column supercritical fluid chromatography of sesquiterpene lactones with different carbon skeletons. *J. Chromatogr. A* 1997, 779, 315–320.

[85] Taylor, L. T., Packed column supercritical fluid chromatography of hydrophilic analytes via water-rich modifiers. *J. Chromatogr. A* 2012, 1250, 196–204.

[86] Pyo, D., Separation of vitamins by supercritical fluid chromatography with water-modified carbon dioxide as the mobile phase. *J. Biochem. Biophys. Methods* 2000, 43, 113–123.

[87] Salvador, A., Herbreteau, B., Lafosse, M., Dreux, M., Subcritical fluid chromatography of monosaccharides and polyols using silica and trimethylsilyl columns. *J. Chromatogr. A* 1997, 785, 195–204.

[88] Suto, K., Kakinuma, S., Ito, Y., Sagara, K., Iwasaki, H., Itokawa, H., Determination of berberine and palmatine in Phellodendri Cortex using ion-pair supercritical fluid chromatography on-line coupled with ion-pair supercritical fluid extraction by on-column trapping. *J. Chromatogr. A* 1997, 786, 371–376.

[89] Lesellier, E., Krstulović, A. M., Tchapla, A., Specific effects of modifiers in subcritical fluid chromatography of carotenoid pigments. *J. Chromatogr. A* 1993, 641, 137–145.

[90] Lesellier, E., Krstulovic, A. M., Tchapla, A., Influence of the modifiers on the nature of the stationary phase and the separation of carotenes in sub-critical fluid chromatography. *Chromatographia* 1993, 36, 275–282.

[91] Lesellier, E., Tchapla, A., Retention behavior of triglycerides in octadecyl packed subcritical fluid chromatography with CO_2/Modifier mobile phases. *Anal. Chem.* 1999, 71, 5372–5378.

[92] Funada, Y., Hirata, Y., Retention behavior of triglycerides in subcritical fluid chromatography with carbon dioxide mobile phase. *J. Chromatogr. A* 1997, 764, 301–307.

[93] Funada, Y., Hirata, Y., Analysis of plant oils by subcritical fluid chromatography using pattern fitting. *J. Chromatogr. A* 1998, 800, 317–325.

[94] Duval, J., Colas, C., Pecher, V., Poujol, M., Tranchant, J.-F., Lesellier, É., Contribution of supercritical fluid chromatography coupled to high resolution mass spectrometry and UV detections for the analysis of a complex vegetable oil – Application for characterization of a Kniphofia uvaria extract. *Comptes Rendus Chim.* 2016, 19, 1113–1123.

[95] Raynor, M. W., Kithinji, J. P., Bartle, K. D., Games, D. E., Mylchreest, I. C., Lafont, R., Morgan, E. D., Wilson, I. D., Packed column supercritical-fluid chromatography and linked super-critical-fluid chromatography-mass spectrometry for the analysis of phytoecdysteroids from Silene nutans and Silene otites. *J. Chromatogr. A* 1989, 467, 292–298.

[96] Dost, K., Davidson, G., Analysis of artemisinin by a packed-column supercritical fluid chromatography-atmospheric pressure chemical ionisation mass spectrometry technique. *Analyst* 2003, 128, 1037–1042.

[97] Sandra, P., Medvedovici, A., Zhao, Y., David, F., Characterization of triglycerides in vegetable oils by silver-ion packed-column supercritical fluid chromatography coupled to mass spectroscopy with atmospheric pressure chemical ionization and coordination ion spray. *J. Chromatogr. A* 2002, 974, 231–241.

[98] Wang, M., Carrell, E. J., Chittiboyina, A. G., Avula, B., Wang, Y.-H., Zhao, J., Parcher, J. F., Khan, I. A., Concurrent supercritical fluid chromatographic analysis of terpene lactones and ginkgolic acids in Ginkgo biloba extracts and dietary supplements. *Anal. Bioanal. Chem.* 2016, 408, 4649–4660.

[99] Fu, Q., Li, Z., Sun, C., Xin, H., Ke, Y., Jin, Y., Liang, X., Rapid and simultaneous analysis of sesquiterpene pyridine alkaloids from Tripterygium wilfordii Hook. f. Using supercritical fluid chromatography-diode array detector-tandem mass spectrometry. *J. Supercrit. Fluids* 2015, 104, 85–93.

[100] Chen, Y., Lehotay, S. J., Moreau, R. A., Supercritical fluid chromatography-tandem mass spectrometry for the analysis of lipid A. *Anal. Methods* 2013, 5, 6864–6869.

[101] Wang, M., Wang, Y.-H., Avula, B., Radwan, M. M., Wanas, A. S., Mehmedic, Z., van Antwerp, J., ElSohly, M. A., Khan, I. A., Quantitative determination of cannabinoids in cannabis and cannabis products using ultra-high-performance supercritical fluid chromatography and diode array/mass spectrometric detection. *J. Forensic Sci.* 2017, 62, 602–611.

[102] Lee, J. W., Uchikata, T., Matsubara, A., Nakamura, T., Fukusaki, E., Bamba, T., Application of supercritical fluid chromatography/mass spectrometry to lipid profiling of soybean. *J. Biosci. Bioeng.* 2012, 113, 262–268.

[103] Jiang, Q., Liu, W., Li, X., Zhang, T., Wang, Y., Liu, X., Detection of related substances in polyene phosphatidyl choline extracted from soybean and in its commercial capsule by comprehensive supercritical fluid chromatography with mass spectrometry compared with HPLC with evaporative light scattering detection. *J. Sep. Sci.* 2016, 39, 350–357.

[104] Prothmann, J., Sun, M., Spégel, P., Sandahl, M., Turner, C., Ultra-high-performance supercritical fluid chromatography with quadrupole-time-of-flight mass spectrometry (UHPSFC/QTOF-MS) for analysis of lignin-derived monomeric compounds in processed lignin samples. *Anal. Bioanal. Chem.* 2017, DOI: 10.1007/s00216-017-0663-5.

[105] Lesellier, E., Valarché, A., West, C., Dreux, M., Effects of selected parameters on the response of the evaporative light scattering detector in supercritical fluid chromatography. *J. Chromatogr. A* 2012, 1250, 220–226.

[106] Morin-Allory, L., Herbreteau, B., High-performance liquid chromatography and supercritical fluid chromatography of monosaccharides and polyols using light-scattering detection: Chemometric studies of the retentions. *J. Chromatogr. A* 1992, 590, 203–213.

[107] Herbreteau, B., Lafosse, M., Morin-Allory, L., Dreux, M., Analysis of sugars by supercritical fluid chromatography using polar packed columns and light-scattering detection. *J. Chromatogr. A* 1990, 505, 299–305.

[108] Thompson, J., Strode III, B., Taylor, L. T., van Beek, T. A., Supercritical fluid chromatography of ginkgolides A, B, C and J and bilobalide. *J. Chromatogr. A* 1996, 738, 115–122.

[109] Samimi, R., Xu, W. Z., Alsharari, Q., Charpentier, P. A., Supercritical fluid chromatography of North American ginseng extract. *J. Supercrit. Fluids* 2014, 86, 115–123.

[110] Yip, S.-H. H., Ashraf-Khorassani, M., Taylor, L. T., Analytical scale supercritical fluid fractionation and identification of single polar lipids from deoiled soybean lecithin. *J. Sep. Sci.* 2008, 31, 1290–1298.

[111] Deschamps, F. S., Lesellier, E., Bleton, J., Baillet, A., Tchapla, A., Chaminade, P., Glycolipid class profiling by packed-column subcritical fluid chromatography. *J. Chromatogr. A* 2004, 1040, 115–121.

[112] Lesellier, E., Gaudin, K., Chaminade, P., Tchapla, A., Baillet, A., Isolation of ceramide fractions from skin sample by subcritical chromatography with packed silica and evaporative light scattering detection. *J. Chromatogr. A* 2003, 1016, 111–121.

[113] Deschamps, F. S., Gaudin, K., Lesellier, E., Tchapla, A., Ferrier, D., Baillel, A., Chaminade, P., Response enhancement for the evaporative light scattering detection for the analysis of lipid classes and molecular species. *Chromatographia* 2001, 54, 607–611.

[114] Li, J.-R., Li, M., Xia, B., Ding, L.-S., Xu, H.-X., Zhou, Y., Efficient optimization of ultra-high-performance supercritical fluid chromatographic separation of Rosa sericea by response surface methodology. *J. Sep. Sci.* 2013, 36, 2114–2120.

[115] Li, J., Quantitative analysis of cosmetics waxes by using supercritical fluid extraction (SFE)/supercritical fluid chromatography (SFC) and multivariate data analysis. *Chemom. Intell. Lab. Syst.* 1999, 45, 385–395.

[116] Brossard, S., Lafosse, M., Dreux, M., Analysis of synthetic mixtures of waxes by supercritical fluid chromatography with packed columns using evaporative light-scattering detection. *J. Chromatogr. A* 1992, 623, 323–328.

[117] Brossard, S., Lafosse, M., Dreux, M., Becart, J., Abnormal composition of commerical waxes revealed by supercritical fluid chromatography. *Chromatographia* 1993, 36, 268–274.

[118] Lesellier, E., Mith, D., Dubrulle, I., Method developments approaches in supercritical fluid chromatography applied to the analysis of cosmetics. *J. Chromatogr. A* 2015, 1423, 158–168.

[119] Zhao, Y., McCauley, J., Pang, X., Kang, L., Yu, H., Zhang, J., Xiong, C., Chen, R., Ma, B., Analytical and semipreparative separation of 25 (R/S)-spirostanol saponin diastereomers using supercritical fluid chromatography. *J. Sep. Sci.* 2013, 36, 3270–3276.

[120] Ashraf-Khorassani, M., Taylor, L. T., Martin, M., Supercritical fluid extraction of Kava lactones from Kava root and their separation via supercritical fluid chromatography. *Chromatographia* 1999, 50, 287–292.

[121] Coleman, W. M., Dube, M. F., Ashraf-Khorassani, M., Taylor, L. T., Isomeric enhancement of davanone from natural davana oil aided by supercritical carbon dioxide. *J. Agric. Food Chem.* 2007, 55, 3037–3043.

[122] Zhang, L., Sun, A., Li, A., Kang, J., Wang, Y., Liu, R., Isolation and purification of osthole and imperatorin from Fructus Cnidii by semi-preparative supercritical fluid chromatography. *J. Liq. Chromatogr. Relat. Technol.* 2017, 40, 407–414.

[123] Yamauchi, Y., Saito, M., Fractionation of lemon-peel oil by semi-preparative supercritical fluid chromatography. *J. Chromatogr. A* 1990, 505, 237–246.

[124] Choo, Y. M., Ma, A. N., Yahaya, H., Yamauchi, Y., Bounoshita, M., Saito, M., Separation of crude palm oil components by semipreparative supercritical fluid chromatography. *J. Am. Oil Chem. Soc.* 1996, 73, 523–525.

[125] Yang, B., Xin, H., Wang, F., Cai, J., Liu, Y., Fu, Q., Jin, Y., Liang, X., Purification of lignans from Fructus Arctii using off-line two-dimensional supercritical fluid chromatography/reversed-phase liquid chromatography. *J. Sep. Sci.* 2017, 40, 3231–3238.

[126] Saito, M., Yamauchi, Y., Isolation of tocopherols from wheat germ oil by recycle semi-preparative supercritical fluid chromatography. *J. Chromatogr. A* 1990, 505, 257–271.

[127] Stevenson, P. G., Tarafder, A., Guiochon, G., Comprehensive two-dimensional chromatography with coupling of reversed phase high performance liquid chromatography and supercritical fluid chromatography. *J. Chromatogr. A* 2012, 1220, 175–178.

[128] Li, K., Fu, Q., Xin, H., Ke, Y., Jin, Y., Liang, X., Alkaloids analysis using off-line two-dimensional supercritical fluid chromatography × ultra-high performance liquid chromatography. *Analyst* 2014, 139, 3577–3587.

[129] Bonaccorsi, I., Cacciola, F., Utczas, M., Inferrera, V., Giuffrida, D., Donato, P., Dugo, P., Mondello, L., Characterization of the pigment fraction in sweet bell peppers (Capsicum annuum L.) harvested at green and overripe yellow and red stages by offline multidimensional convergence chromatography/liquid chromatography–mass spectrometry. *J. Sep. Sci.* 2016, 39, 3281–3291.

[130] François, I., Pereira, A. dos S., Sandra, P., Considerations on comprehensive and off-line supercritical fluid chromatography×reversed-phase liquid chromatography for the analysis of triacylglycerols in fish oil. *J. Sep. Sci.* 2010, 33, 1504–1512.

[131] Hirata, Y., Hashiguchi, T., Kawata, E., Development of comprehensive two-dimensional packed column supercritical fluid chromatography. *J. Sep. Sci.* 2003, 26, 531–535.

[132] Hirata, Y., Sogabe, I., Separation of fatty acid methyl esters by comprehensive two-dimensional supercritical fluid chromatography with packed columns and programming of sampling duration. *Anal. Bioanal. Chem.* 2004, 378, 1999–2003.

[133] François, I., dos Santos Pereira, A., Lynen, F., Sandra, P., Construction of a new interface for comprehensive supercritical fluid chromatography×reversed phase liquid chromatography (SFC×RPLC). *J. Sep. Sci.* 2008, 31, 3473–3478.

[134] François, I., Sandra, K., Sandra, P., Comprehensive liquid chromatography: Fundamental aspects and practical considerations–A review. *Anal. Chim. Acta* 2009, 641, 14–31.

[135] Taguchi, K., Fukusaki, E., Bamba, T., Simultaneous analysis for water- and fat-soluble vitamins by a novel single chromatography technique unifying supercritical fluid chromatography and liquid chromatography. *J. Chromatogr. A* 2014, 1362, 270–277.

[136] Bamba, T., Lee, J. W., Matsubara, A., Fukusaki, E., Metabolic profiling of lipids by supercritical fluid chromatography/mass spectrometry. *J. Chromatogr. A* 2012, 1250, 212–219.

[137] Grand-Guillaume Perrenoud, A., Guillarme, D., Boccard, J., Veuthey, J.-L., Barron, D., Moco, S., Ultra-high performance supercritical fluid chromatography coupled with quadrupole-time-of-flight mass spectrometry as a performing tool for bioactive analysis. *J. Chromatogr. A* 2016, 1450, 101–111.

[138] Jones, M. D., Avula, B., Wang, Y.-H., Lu, L., Zhao, J., Avonto, C., Isaac,v G., Meeker, L., Yu, K., Legido-Quigley, C., Smith, N., Khan, I. A., Investigating sub-2 μm particle stationary phase

supercritical fluid chromatography coupled to mass spectrometry for chemical profiling of chamomile extracts. *Anal. Chim. Acta* 2014, 847, 61–72.

[139] Duval, J., Pecher, V., Poujol, M., Tranchant, J.-F., Lesellier, E., Selective supercritical fluid extraction (SFE) coupled to impregnation of solid matrix for the cosmetic formulation. *Planta Med.* 2016, 82, S1–S381.

Chandan L. Barhate, Erin E. Jordan, Philip A. Searle

8 Analytical chiral supercritical fluid chromatography

Abstract: Chirality is an important property of pharmaceutical drugs, since each enantiomer may exhibit marked differences in pharmacology, toxicology, pharmacokinetics, and metabolism. Supercritical fluid chromatography (SFC) has proven to be an essential tool for the analysis of chiral compounds. This chapter describes the importance of chirality in the pharmaceutical industry and summarizes analytical SFC instrumentation. Commercially available chiral stationary phases are reviewed, along with selection of mobile phase and the use of tandem column chromatography. Example applications of the use of chiral analytical SFC for the medicinal/synthetic chemist have also been described.

Keywords: supercritical fluid chromatography, chiral separations, chiral stationary phases, pharmaceutical applications of chiral supercritical fluid chromatography

8.1 Introduction to chirality

In the field of chemistry, chirality is a molecular property wherein mirror images of a molecule cannot be superimposed. Two structural isomers, having the same constitution and molecular formula, can differ in the spatial arrangement of atoms within the molecule, giving rise to stereoisomers. Most commonly in organic chemistry, chirality is the result of four different substituents connected to a central carbon atom giving rise to an asymmetric carbon atom (chiral center). The mirror image stereoisomers are referred to as enantiomers or optical isomers (Figure 8.1).

Where more than one chiral center exists in a molecule then the number of possible stereoisomers is increased to 2^N, where N is the number of chiral centers present. For two chiral centers, the four possible stereoisomers exist as two pairs of enantiomers with each pair being, by definition, nonsuperimposable mirror images having opposite configuration at both chiral centers, and four pairs of diastereomers with each pair differing by the configuration at one chiral center (Figure 8.2).

The configuration of each asymmetric carbon center is labeled (R) or (S), based upon the priority of the attached substituents. Diastereoisomers may differ in physical–chemical properties such as melting point, density, and chemical

Chandan L. Barhate, Erin E. Jordan, Philip A. Searle, Discovery Chemistry and Technology, AbbVie Inc., North Chicago
Philip A. Searle

https://doi.org/10.1515/9783110500776-008

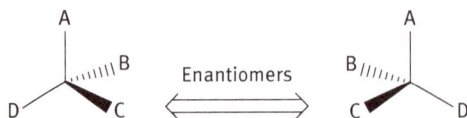

Figure 8.1: Stereoisomers with one chiral center.

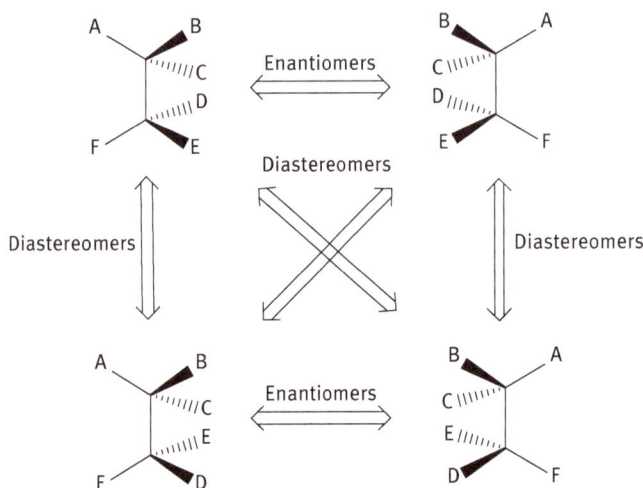

Figure 8.2: Stereoisomers with two chiral centers.

reactivity, whereas enantiomers share identical physical–chemical properties but can rotate plane-polarized light in equal but opposite directions and are hence termed optically active. A 1:1 mixture of enantiomers is termed racemic and produces no net rotation of plane-polarized light. Nonequal mixtures of enantiomers are termed scalemic and will have reduced optical activity compared to the pure single enantiomer.

In many applications, the ability to prepare a single, optically pure, enantiomer of a compound is of paramount importance and requires the capability for accurate analysis to determine enantiomeric purity. Single enantiomers of a compound can be prepared by stereoselective chemical synthesis, but their preparation may also frequently require a separation step, such as enantioselective crystallization, resolution of diastereomeric salt mixtures, enzymatic resolution, or physical–chemical separation of diastereomeric derivatives. Chromatographic techniques including high-performance liquid chromatography (HPLC), supercritical fluid chromatography (SFC), or simulated moving bed chromatography (SMB) can use a chiral stationary phase (CSP) to enable separation and analysis of enantiomers.

8.2 Importance of chirality in the pharmaceutical industry

Enantiomers of racemic drugs may exhibit marked differences in pharmacology, toxicology, pharmacokinetics, and metabolism. The human body, with its numerous homochiral compounds, behaves as a chiral selector, interacting with and metabolizing enantiomers through different biological pathways [1, 2]. In some cases, one enantiomer may display higher levels of activity and quantitative potency, while the other can be potentially toxic. In the pharmaceutical industry, it has become increasingly important to perform chiral separation and analysis of racemic drugs to find optimal and safe therapies for the patient.

The classic example of the impact different enantiomers of a racemic drug can have is the case of thalidomide, originally introduced as a nonbarbiturate sedative with antinausea properties. Thalidomide was prescribed as a morning sickness treatment for pregnant women, but was removed from the market in the early 1960s due to severe teratogenic effects (birth defects). In all medical treatments, thalidomide, with its single chiral center, was administered as the racemate (Figure 8.3). However, the two enantiomers have very different pharmacological activities, with the (R) enantiomer showing biological activity. While the (S) enantiomer was originally thought to be inactive, its presence in the body produced the adverse toxic effect. The problem was further complicated by the fact that the enantiomers of thalidomide are subject to chiral inversion in vitro, with the (R) enantiomer undergoing chiral inversion more quickly than the (S) enantiomer [3].

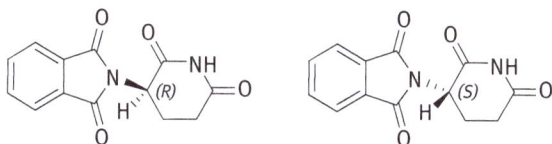

Figure 8.3: Stereoselectivity of thalidomide enantiomers.

Many pharmaceutical drugs are chiral, like thalidomide, and often one enantiomer displays targeted biological activity. In these cases, it is critical to distinguish the single enantiomer from the racemic form for biological testing as early as possible, as they may differ in efficacies, dosage, side product profiles, or toxicities. Furthermore, the decision to use a single enantiomer versus a mixture of enantiomers of a particular drug is much more straightforward when it can be made before the drug goes into clinical trials.

Several commercially available pharmaceutical drugs are marketed as racemic mixtures. For example, β-blockers intended for the treatment of hypertension, such as acebutolol (Sectral), metoprolol (Lopressor), and pindolol (Visken), are typically marketed as racemates. However, for certain patients suffering from both hypertension and hyperthyroidism, β-blocking drugs cannot be administered because one enantiomer may inhibit the conversion of thyroxin (T4) to triiodothyronine (T3). In the case of propranolol (Hemangeol), it has been demonstrated that both racemic D,L-propranolol and D-propranolol cause the inhibition of T4 to T3; however, the D-propranolol can be administered as a specific, single drug for the treatment of hypertension without the adverse effect [2, 4].

Calcium channel antagonists are another example of racemic therapies, with drugs such as nisoldipine (Sular), felodipine (Plendil), and manidipine (Manyper) also marketed as antihypertension drugs. Verapamil (generic) is sold both as a racemate and single enantiomer, $S(-)$ verapamil, for treatment of hypertension, as the $S(-)$ enantiomer has shown a 10–20-fold increase in potency. However, $R(+)$ verapamil has another possible application in cancer chemotherapy and could be used as a multidrug resistance for patients requiring cardio- and chemotherapy, since the low potency of $R(+)$ verapamil allows it to be prescribed in higher doses without adverse cardiotoxicity effect [5].

Of the top 200 pharmaceutical products by retail sales in 2016, approximately 65% of the small-molecule drugs contain at least one chiral center [6]. Since drugs that are produced by chemical synthesis are often racemic, or produced by stereoselective synthesis with less than 100% stereoselectivity, there is a substantial need to separate and quantify synthetic products in the pharmaceutical industry [7].

8.3 Introduction to SFC

Six decades after the original proposal by James Lovelock to employ high liquid-like density mobile phases with very low gas-like viscosities [8], SFC has emerged as a powerful separation technique. During the 1980s it was believed that SFC packed columns could not produce more than ≈20,000 plates [9]. In 1990, Berger and Wilson [10] showed that it was possible to obtain very high efficiencies with SFC and achieved 220,000 plates in 4.6 mm i.d., 2.2 m long column packed with 5 μm particles. In 1991, Cui and Olesik [11] demonstrated the use of high concentration modifiers in liquefied CO_2 and observed that the mobile phase was not a supercritical fluid and termed it "enhanced-fluidity mobile phase."

The preference for SFC, especially in the pharmaceutical industry, arises from the exclusive usage of compressed carbon dioxide as an eluent [12–17]. Carbon dioxide in its sub/super critical state (>7 MPa and above 31 °C) solubilizes a large number of polar and nonpolar compounds in the presence of a small amount of organic solvents [17–20], and addition of modifiers such as methanol, along with buffers, is often used to increase the elution strength of pure supercritical CO_2.

SFC has been widely used for chiral separations, traditionally using 5 μm particles functionalized with a chiral selector, packed in 15–25 cm long columns, providing separations in the 10–40 min timeframe. Small particle size columns can be used to obtain analogous resolutions with shorter columns due to the increased chromatographic efficiency. However, since column backpressure is inversely proportional to the square of particle diameter, historically, separation scientists were limited by the upper pressure limits of conventional HPLC instrumentation. Due to the lower viscosity of the mobile phases used in SFC, conventional instrumentation could be effectively utilized with smaller particle size columns [15, 18, 20, 21].

Recently, analytical SFC has been applied for quantitative analysis using optimized conditions and the method validated [22]. A comparison between ultra-HPLC and SFC for impurity profiling was demonstrated by Wang et al. in 2011 [23]. The authors developed an orthogonal chromatographic method on SFC for the analysis of mometasone furoate and its impurities, and this method was partially validated.

Chromatographic resolution of closely related compounds is a challenging task in pharmaceutical analysis [24, 25]. Regalado et al. [26] showed that achiral and chiral SFC can be effectively utilized for separation of dehalogenated impurities from active pharmaceutical ingredients (API). In 2016, two-dimensional liquid chromatography (LC)-SFC system (heart-cutting mode) was demonstrated for simultaneous quantitative achiral and chiral analysis by Venkatramani et al. The authors showed that achiral purity results could be obtained in the first dimension by reverse-phase liquid chromatography (RPLC) and enantiomer separation with purity information in the second dimension by SFC [27].

8.4 Analytical SFC instrumentation

The instrumentation of modern SFC has been described in detail in a recent publication by Berger [28]. Herein, we summarize important aspects of SFC instrumentation and its idiosyncrasies:

SFC instrumentation is very similar to that of HPLC, except for the CO_2 pumping system and backpressure regulator (BPR) (Figure 8.4). SFC-grade CO_2 cylinders, with a full-length educator tube, typically at a pressure of 55 bar at 20 °C, are utilized for CO_2 delivery to the system. Precompression of CO_2 by a booster pump is required before it enters the metering pump to ensure flow uniformity, to eliminate the pressure fluctuations commonly caused in pumps that both compress and meter the flow. Column temperature in SFC is controlled by a thermostated column compartment, typically between 10 °C and 90 °C, enabling both cooling and heating.

One of the advantages of SFC is that it can utilize both gas chromatography and LC-like detectors. When pure CO_2 is used as a mobile phase, flame ionization detection (FID), electron capture detection, chemiluminescence, and other detection approaches are possible [18]. Although FID-like detectors are limited to using only a mobile phase consisting of pure CO_2 and modifier gradients are not possible, pressure gradients can be used for compound elution. Other detectors that can be used in SFC include UV-VIS diode array detectors, evaporative light scattering detectors (ELSD), charged aerosol detectors, and mass spectrometry (MS). With a UV-Vis detector it is important to ensure that it can withstand the maximum pressure generated by the pump and BPR. The BPR (usually set above 8 MPa) is required to maintain mobile phase in a compressed state and BPR pressure and temperature settings are

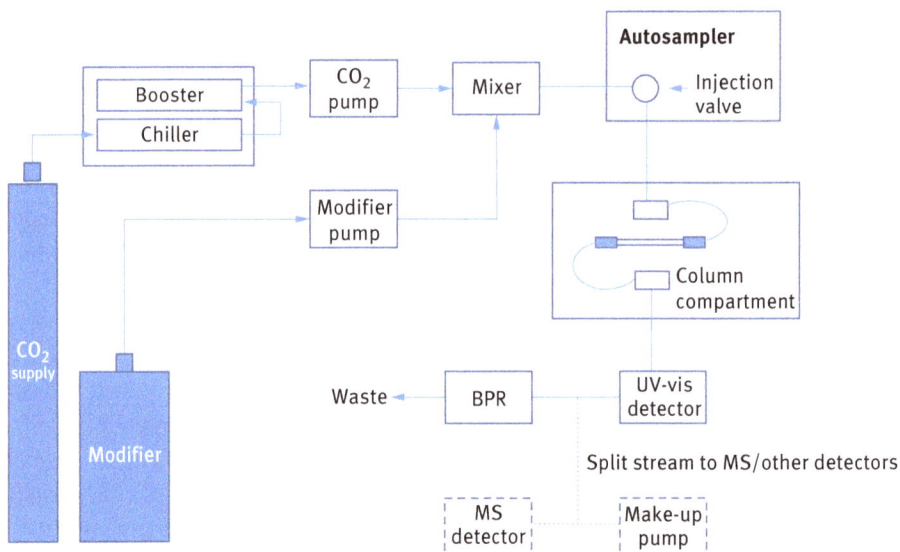

Figure 8.4: Analytical SFC instrumentation.

adjustable. For other detectors such as MS or ELSD the flow must be split before the BPR and the split flow introduced to the detector, often using a make-up pump.

8.5 Chiral stationary phases for analytical SFC

More than 100 CSPs are commercially available [29]; among them phenylcarbamates of polysaccharides (including amylose and cellulose) have been recognized as the most effective for the resolution of a wide variety of chiral compounds [30]. The separation of a pair of enantiomers is predominantly the result of stereoselective interactions with the CSP [29, 35]. Unfortunately, because these interactions are unknown, it is very difficult to predict separation and elution order from the chemical structure of the analyte alone; and column screening is often necessary. CSPs for SFC are commercially available in four major categories: polysaccharide, immobilized polysaccharide, brush or Pirkle-type, and fluorinated phases, as well as some specialized phases such as macrocyclic glycopeptide (teicoplanin [T] and teicoplanin aglycone [TAG]) and cyclofructan-based (LARIHC CF6P).

8.5.1 Polysaccharide CSP

Classically coated polysaccharides on silica, based on either cellulose or amylose, have been used as chiral supports in HPLC and SFC (Tables 8.1 and 8.2) since the 1970s [36–39]. The interaction between the analyte and the polymer CSPs is directed

Table 8.1: Coated polysaccharide cellulose derivative CSPs.

CHIRALCEL OD, Lux Cellulose-1, RegisCell™, Chromegachiral CCO	**CHIRALCEL OA**
Tris(3,5-dimethylphenylcarbamate)	Triacetate
CHIRALCEL OB	**CHIRALCEL OC**
Tribenzoate	Tris(phenylcarbamate)
CHIRALCEL OF	**CHIRALCEL OG**
Tris(4-chlorophenylcarbamate)	Tris(4-methylphenylcarbamate)
CHIRALCEL OJ, Lux Cellulose-3	**CHIRALCEL OK**
Tris(4-methylbenzoate)	Tricinnamate
CHIRALCEL OX Lux Cellulose-4	**CHIRALCEL OZ, Lux Cellulose-2,**
Tris(4-chloro-3-methylphenylcarbamate)	Tris(3-chloro-4-methylphenylcarbamate)

by one or a combination of mechanisms, including hydrogen bonding, π–π interaction, and/or dipole stacking and inclusion complexes [40, 41].

Phenylcarbamates are another widely used small-molecule substituent class of polysaccharide phases. To prepare these phases, cellulose and amylose are converted to various phenylcarbamate derivatives by reacting with phenyl isocyanates [42, 43]. The phenylcarbamate stationary phases can also be tuned to support greater hydrogen bonding interactions with the racemate by increasing the strength of the

Table 8.2: Coated polysaccharide amylose derivative CSPs.

CHIRALPAK AD, Lux Amylose-1, RegisPack™, ChromegaChiral CCA	CHIRALPAK AS, ChromegaChiral CCS

CHIRALPAK AD, Lux Amylose-1, RegisPack™, ChromegaChiral CCA

Tris(3,5-dimethylphenylcarbamate)

CHIRALPAK AS, ChromegaChiral CCS

Tris((S)-α-methylbenzylcarbamate)

CHIRALPAK AY, Lux Amylose-2, RegisPack™ CLA-1,

Tris(5-chloro-2-methylphenylcarbamate)

CHIRALPAK AZ

Tris(3-chloro-4-methylphenylcarbamate)

electron-donating substituents on the phenyl group and, therefore, the electron density at the carbonyl oxygen [30].

An additional mechanism that supports the broad applicability of polysaccharide CSPs is the conformational position of the substituents on the phenyl group [34, 42–45]. With the addition of an electron-donating or electron-withdrawing group at the ortho position of a cellulose-based CSP, an increase in chiral recognition is observed [43]. Conversely, similar modifications to an amylose-based CSP have shown a diminished effect for some racemates, suggesting that the interaction between the racemate and CSP is directly related to the conformational interplay between both moieties [30].

8.5.2 Immobilized CSP

An immobilized CSP combines the benefits of a polysaccharide-based chiral selector with the advantages of the immobilization process, which leads to a wider range of solvent compatibility [46–50]. Solvents of medium polarity, such as tetrahydrofuran, ethyl acetate, chloroform, and methylene chloride can be explored on immobilized CSPs, allowing a much broader enantioselective capacity [51]. The chiral polymers incorporated in CHIRALPAK IA and CHIRALPAK IB CSPs are related to the polymers used in the coated equivalent phases: tris(3,5-dimethylphenylcarbamate) of amylose for CHIRALPAK IA and tris(3,5-dimethylphenylcarbamate) of cellulose for CHIRALPAK IB (Tables 8.3 and 8.4) [46, 50]. CHIRALPAK

IC, based on immobilized tris(3,5-dichlorophenylcarbamate) of cellulose, has no coated analogue.

Table 8.3: Immobilized polysaccharide cellulose derivative CSPs.

CHIRALPAK IB	CHIRALPAK IC, Lux i-Cellulose-5
Tris(3,5-dimethylphenylcarbamate)	Tris(3,5-dichlorolphenylcarbamate)

ChromegaChiralCCC

Tris(3,5-dichlorolphenylcarbamate)
and

Tris(3-chloro-4-methylphenylcarbamate)

8.5.3 Brush-type or Pirkle-type CSPs

Brush-type or Pirkle-type CSPs (Table 8.5.) were developed in 1980 by Professor William Pirkle in collaboration with Regis Technologies, which remains the commercial vendor for Pirkle-type columns. Brush-type CSPs form complexes with the analyte through attractive interactions, such as hydrogen bonding, $\pi - \pi$ interactions, dipole-dipole interactions, and minimization of repulsive (steric) interactions [52, 53]. Pirkle-type CSPs have the additional advantage of being commercially available with the chiral selector in two opposite configurations, for example, (R,R) versus (S,S) absolute configuration. The most widely used brush-type or Pirkle columns are the Whelk-O series, bearing both π-acidic and π-basic interaction sites [52, 54]. Terfloth et al. [55] and Pirkle [56] modified the Whelk-O-1 stationary phases and their

Table 8.4: Immobilized polysaccharide amylose derivative CSPs

CHIRALPAK IA, Lux i-Amylose-1	CHIRALPAK ID
Tris(3,5-dimethylphenylcarbamate)	Tris(3-chlorolphenylcarbamate)
CHIRALPAK IE	CHIRALPAK IF
Tris(3,5-dichlorolphenylcarbamate)	Tris(3-chloro-4-methylphenyl-carbamate)
CHIRALPAK IG	
Tris(3,5-dimethylphenylcarbamate)	

analogues by incorporating them onto polysiloxanes, coating them onto silica, and thermally immobilizing them, resulting in a robust immobilized-like column compatible with a wide variety of solvents.

8.5.4 Fluorinated CSP

Fluorinated chiral polysaccharide stationary phases are the latest commercially available CPSs for SFC and provide alternative retention and selectivity to classically coated polysaccharide CSPs (Table 8.6). The addition of a fluorine atom into a phenyl carbamate cellulose structure has been shown to be useful in promoting a fluorophilic retention mechanism, which can provide improved retention for fluorinated compounds [57]. The incorporation of fluorinated groups within biologically active molecules in the pharmaceutical industry has led to the development of these phases to tackle the unique challenges of separating fluorinated chiral analytes [57].

Table 8.5: Pirkle-type CSPs

Whelk-O® 1	Whelk-O® 2
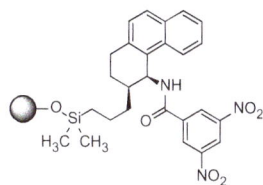	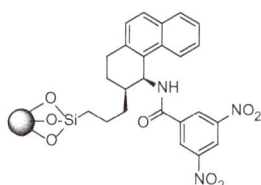
(*R,R*) or (*S,S*)	(*R,R*) or (*S,S*)
Leucine	Phenylglycine
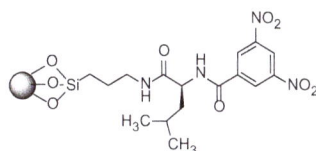	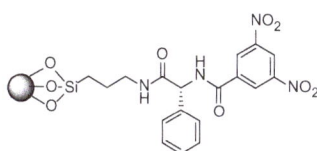
D-Leucine or L-leucine	D-Phenylglycine or L-phenylglycine
β-Gem 1	α-Burke 2
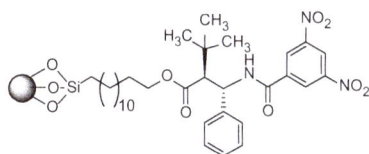	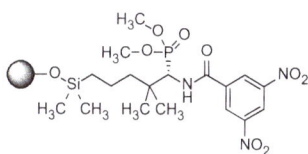
(*R,R*) or (*S,S*)	(*R*) or (*S*)
Pirkle 1-J	ULMO
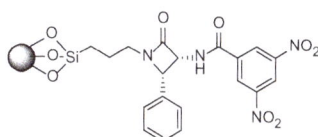	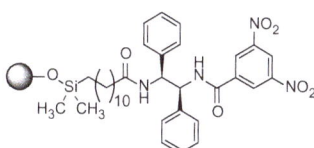
(3*R*,4*S*) or (3*S*, 4*R*)	(*R,R*) or (*S,S*)
DACH-DNB	
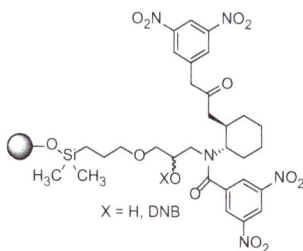	
(*R,R*) or (*S,S*)	

Table 8.6: Fluorinated CSPs.

ChromegaChiral CCO F2	ChromegaChiral CCO F4
tris(2-fluoro-5-methylphenylcarbamate)	tris(4-fluoro-3-methylphenylcarbamate)

8.5.5 Other CSPs

Armstrong et al. [58] introduced macrocyclic glycopeptide CSPs (Table 8.7) in 1994. Teicoplanin and TAG can separate wide classes of compounds by SFC [19, 59]. Cyclofructan CSPs (Table 8.8) have been shown to macrocyclic oligosaccharides consisting of six or more β(2 → 1)-linked D-fructofuranose units. While native cyclofructans showed limited enantioselectivity in early studies, when derivatized with aliphatic or aromatic functional groups, their significant potential as chiral selectors for different classes of compounds was realized [30, 60].

Table 8.7: Glycopeptide CSPs

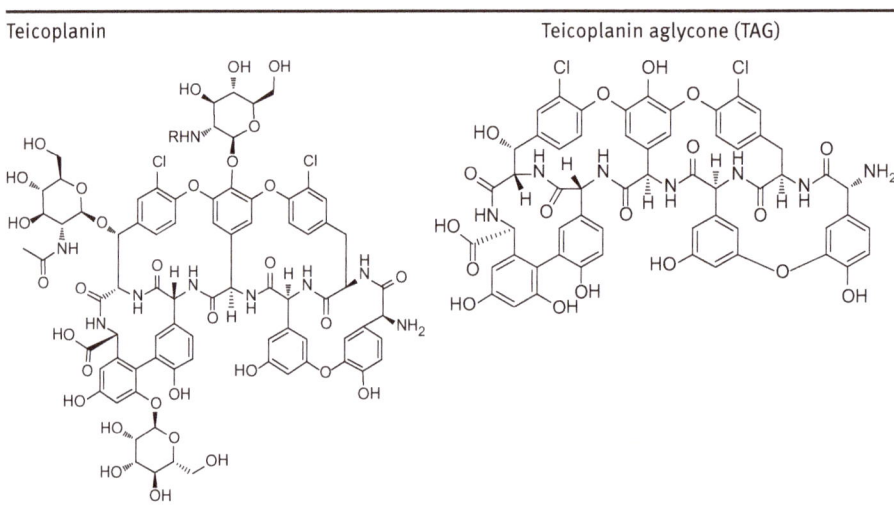

Teicoplanin	Teicoplanin aglycone (TAG)

Table 8.8: Cyclofructan-based CSPs.

Cyclofructan-based chiral selectors

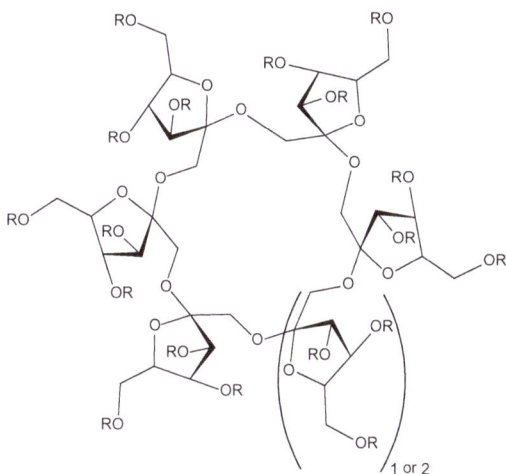

1 or 2

R=H or derivatization group

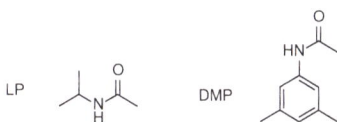

LP

DMP

8.6 Mobile phases for chiral analytical SFC

Supercritical CO_2 is a nonprotic solvent, exhibiting a low dielectric constant; however, as an exclusive mobile phase, supercritical CO_2 is too nonpolar to separate the majority of polar analytes [61]. Klesper and Hartmann [62] reported one of the earliest uses of using liquid polar organic solvents as cosolvents to increase the mobile phase polarity. The use of polar liquid modifiers with CO_2 allows the chromatographer the ability to tune selectivity through different interactions, and the presence of modifiers in the mobile phase has been shown to increase the solvating properties of supercritical CO_2, through the study of partition coefficients between the two phases [63].

Typical solvents used as polar modifiers in analytical SFC are alcohols such as methanol, ethanol, and isopropanol [64]. Acetonitrile, while a good solubilizing solvent, typically acts as a weaker eluent than alcohols toward hydrogen bond donor analytes, and is, therefore, less frequently used [65]. Methanol is the most common and most polar modifier used by chromatographers, accounting for 75% of the usage, mostly due to its complete miscibility with carbon dioxide over a wide range of temperatures and pressures [64–66]. Other organic solvents such as dichloromethane, chloroform, acetone, ethyl acetate, toluene, and tetrahydrofuran can be used when the analytical SFC instrument is equipped with immobilized or Pirkle-type columns, and their neat or blended use has been found to be useful in the separation of drug candidates which are poorly soluble in traditional cosolvents [67]. Additionally, within the last 5 years, the use of blended mobile phases has been studied at leading pharmaceutical companies such as Amgen, Merck, and Pfizer as a way to reduce cosolvent screening through the use of a single mobile phase "cocktail" with a higher hit rate than a single neat solvent [68–71].

Additives are able to decrease ionic interaction with the stationary phase and/or enhance the mobile phase solvating power, resulting in shorter retention times and improved peak shape [39, 72–74]. Typical additives used in chiral analytical SFC to improve peak shape include strong acids such as formic acid, trifluoroacetic acid, and citric acid; bases such as diethylamine, trimethylamine, and ammonia or ammonium acetate; ionic additives such as ammonium hydroxide; and water [38].

8.7 Tandem column analytical SFC

In the analysis of compounds containing more than one chiral center, the chromatographic separation of all possible stereoisomers cannot always be achieved using only one CSP. Where stereoisomers exhibit different selectivity on another CSP, a stepwise approach may be performed, where multiple analytical runs are performed using different chiral columns at each step. However, identification of each component and comparison between each analytical run are not trivial. Another approach is to use two columns in tandem, that is, two individual columns connected together in series [75–78]. This is illustrated schematically in Figure 8.5.

In a tandem column system the overall selectivity is determined by the retention time weighted average of selectivity from each column in series [77]. The mobile phase used and chemistry of each CSP obviously have the most significant impact on combined selectivity, but other parameters can be modified to tune the overall separation. Changing the column lengths or internal diameters can be used to vary the retention time on each column [79]. In SFC, increasing the system backpressure has the effect of decreasing retention time but does not have a marked effect on selectivity [76]. However, in tandem SFC, changing the system backpressure will alter the portion that each column contributes to retention time, and thus can impact the overall selectivity. The order that the columns are connected in series will also have an effect on the separation, since the first column in series experiences a backpressure due to both the second column in series and the instrument BPR. If the column positions are reversed, then each column experiences a significantly different backpressure with the resultant change in retention time contribution.

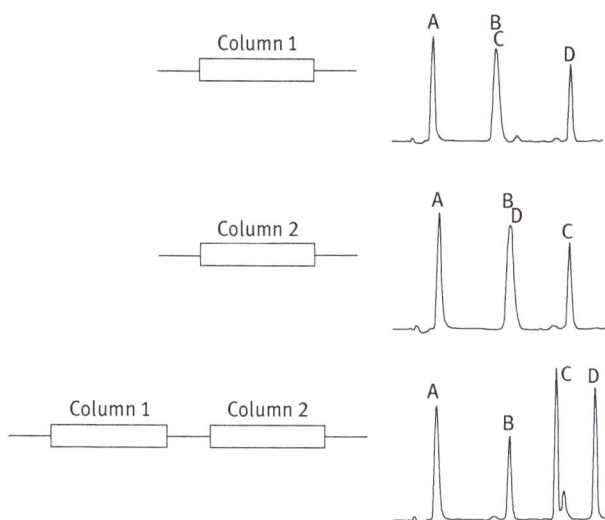

Figure 8.5: Schematic representation of tandem column chiral SFC to separate four stereoisomers (A,B,C,D).

Wang et al. [77] have demonstrated simulation of tandem column SFC based on measurement of retention time on each column separately, allowing accurate prediction of the separation due to the effects of column order and backpressure modification.

8.8 Examples of chiral analytical SFC applications

In the area of drug discovery, analytical chiral SFC has become an established tool for the medicinal chemist in the pursuit of novel therapeutic agents. The need to synthesize drug candidates of greater structural diversity has driven the requirement for the routine analysis and separation of chiral compounds. Two examples from our own laboratories are described below [80, 81].

During the synthesis of a target molecule, the medicinal chemist will often need to develop stereoselective reactions to preferentially synthesize the desired stereoisomer. Rapid analysis of crude reaction mixtures to determine enantiomeric and/or diastereomeric ratios is essential and, in our laboratories, is provided by "walk-up" ("open-access") analytical chiral SFC. These systems are routinely used for analysis of reaction mixtures and also for the analysis of purified products. An Agilent analytical SFC system equipped with an autosampler, a six-position column switcher, CO_2 and modifier pumps, and a UV diodearray detector is configured with standard methods and is the most commonly used chiral column. A custom sample log-in wizard enables the user to rapidly enter sample information and select the columns/

method for analysis (Figure 8.6). Upon completion, the log-in software automatically creates the sample list needed for the multiple analytical runs (Figure 8.7) and coordinates column switching, column equilibration, and sending of a standard reports to the submitter via email.

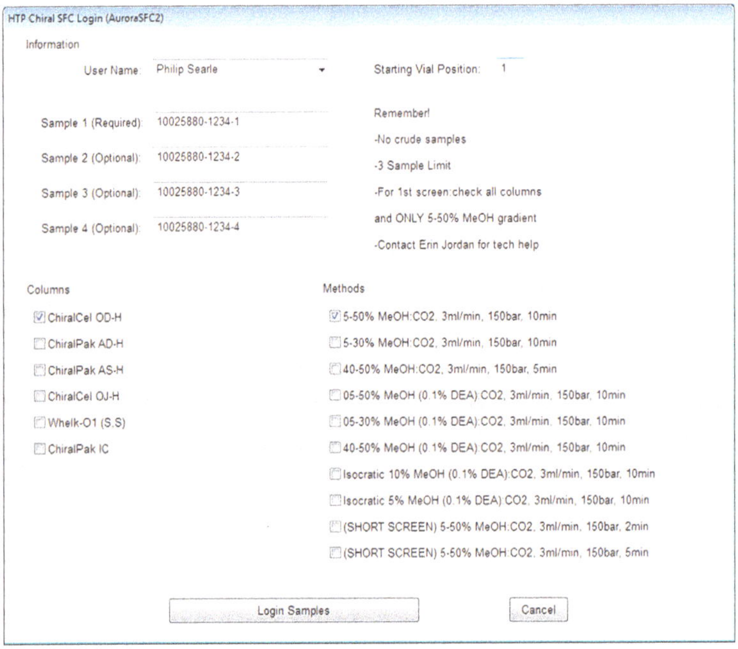

Figure 8.6: Sample log-in for analysis of reaction stereoselectivity optimization.

For more challenging chiral separations, or where the optimum column chemistry and chromatographic method must be determined prior to a preparative-scale SFC separation, a more comprehensive screening system is employed. A similar analytical SFC system to that previously described is enhanced with the addition of two 10-position column switching valves to allow selection of 19 possible columns of different stationary phase chemistry. The first valve allows switching to nine possible column positions, while switching to the 10th bypass position switches flow to the second valve to allow selection of any of the 10 possible columns (Figure 8.8). Both valves are controlled via the custom log-in software and the user only needs to select the column(s) of interest during sample log-in (Figure 8.9) to construct the sample list needed to screen all selected columns (Figure 8.10).

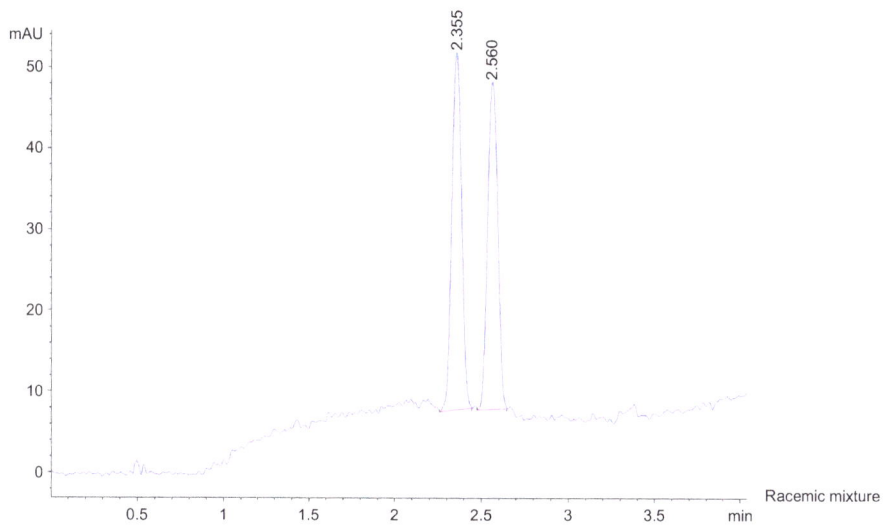

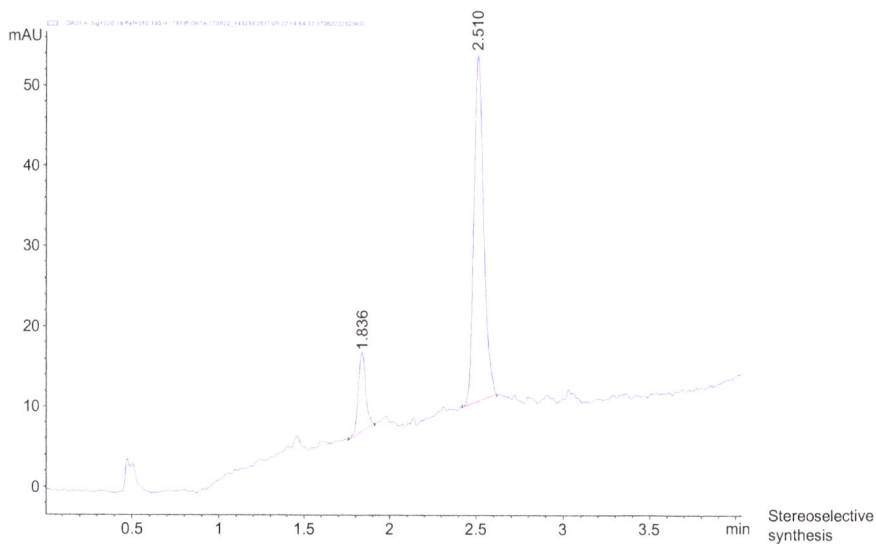

Figure 8.7: SFC analysis of a reaction mixture during optimization of stereoselectivity.
Note: CHIRALPAK® AS-H, 4.6 mm ×100 mm, 5 μm, 5–50% methanol–CO2 at 3 mL/min, 150 bar.

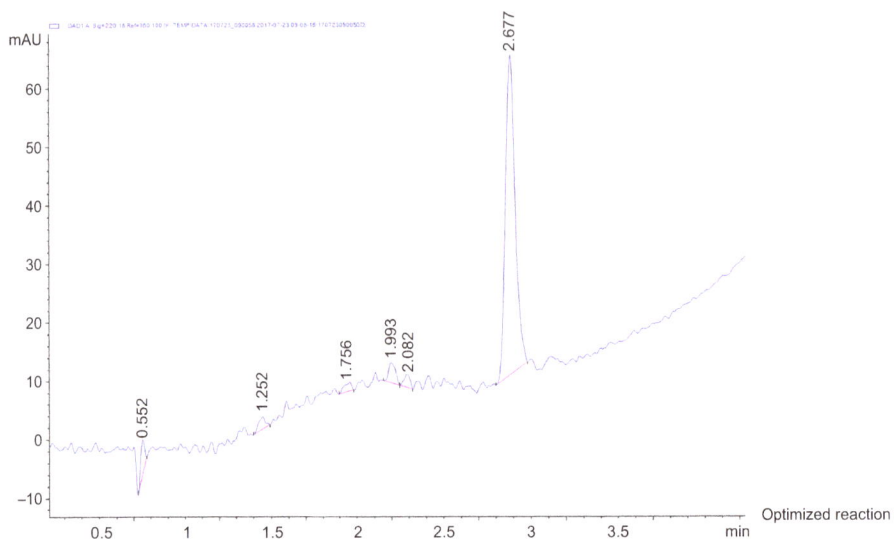

Figure 8.7: (Continued)

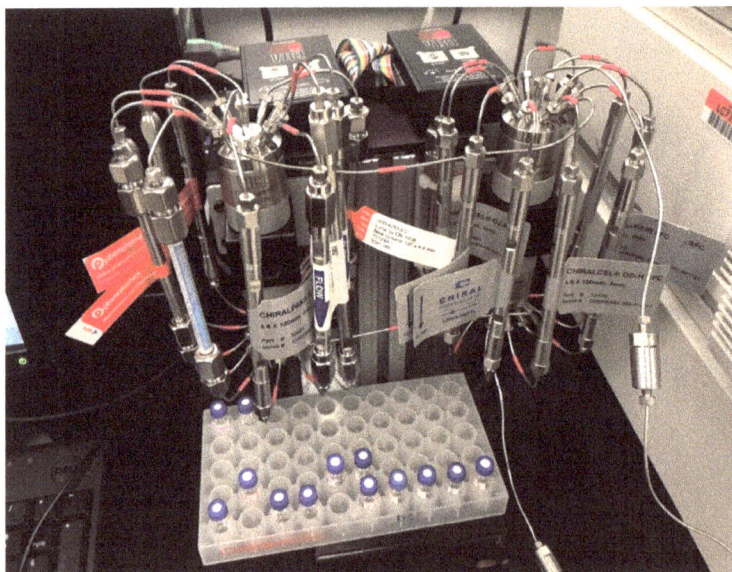

Figure 8.8: Column switching valves for chiral analytical SFC screening system.

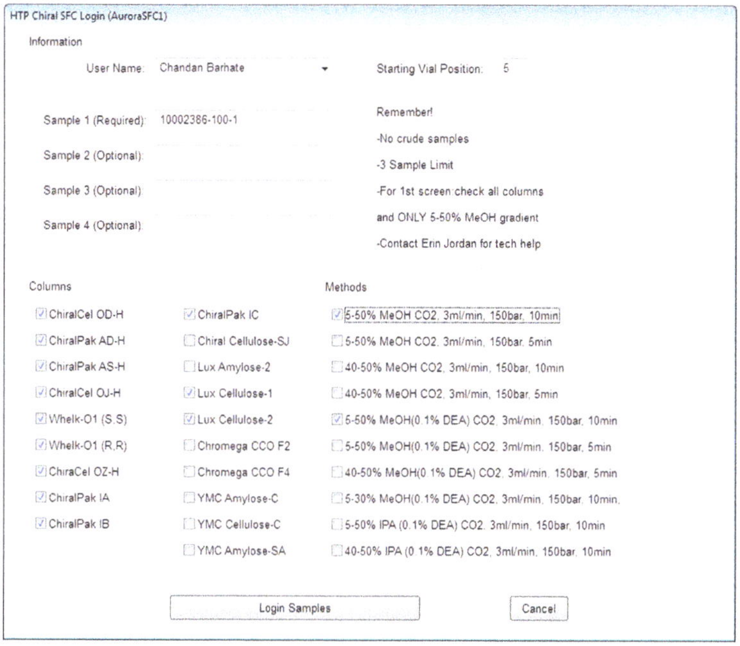

Figure 8.9: Log-in for chiral screen.

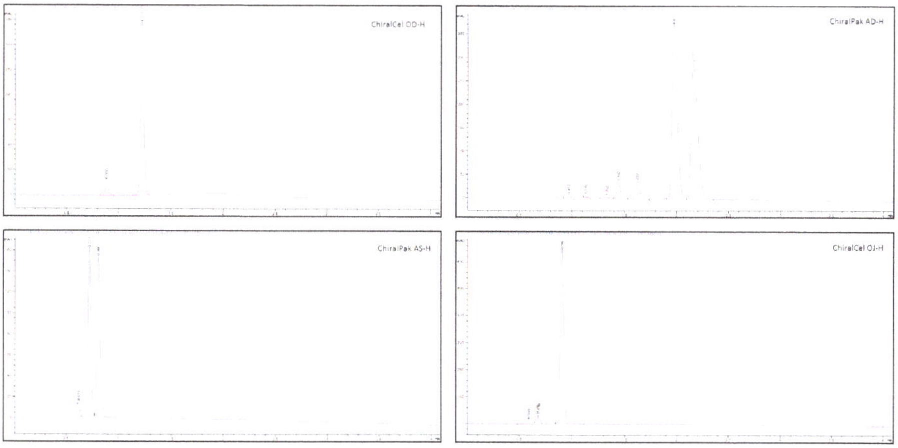

Figure 8.10: Chiral SFC screening results.
Note: Column 4.6 mm × 100 mm, 5 μm, 5–50% methanol–CO_2 at 3 mL/min, 150 bar.

Figure 8.10: (Continued)

8.9 Summary

SFC has proven to be an essential tool for the analysis of chiral compounds. Improvements in instrumentation, chiral stationary phases, and mobile phases continue to push the field forward. The trend toward high-throughput and ultrafast SFC analyses enables chiral screening and reaction monitoring to be accomplished rapidly. With the drive toward drug molecules of increasing chemical diversity and stereochemical complexity, the need for analytical chiral SFC will continue.

References

[1] Landoni, M. and A. Soraci, Pharmacology of chiral compounds 2-Arylpropionic acid derivatives. *Current Drug Metabolism*, 2001. 2(1): pp. 37–51.

[2] Nguyen, L.A., H. He, and C. Pham-Huy, Chiral drugs: an overview. *International Journal of Biomedical Science: IJBS*, 2006. 2(2): pp. 85–100.

[3] Reist, M., et al., Chiral inversion and hydrolysis of thalidomide: mechanisms and catalysis by bases and serum albumin, and chiral stability of teratogenic metabolites. *Chemical Research in Toxicology*, 1998. 11(12): pp. 1521–8.

[4] Harrower, A.D. et al., Thyroxine and triiodothyronine levels in hyperthyroid patients during treatment with propranolol. *Clinical Endocrinology* (Oxford), 1977. 7(1): pp. 41–4.

[5] Satoh, K., T. Yanagisawa, and N. Taira, Coronary vasodilator and cardiac effects of optical isomers of verapamil in the dog. *Journal of Cardiovascular Pharmacology*, 1980. 2(3): pp. 309–18.

[6] Webster, G.T., et al., Resonance Raman spectroscopy can detect structural changes in haemozoin (malaria pigment) following incubation with chloroquine in infected erythrocytes. *Febs Letters*, 2008. 582(7): pp. 1087–1092.

[7] McConathy, J. and M.J. Owens, *Stereochemistry in drug action.* Prim Care Companion. *The Journal of Clinical Psychiatry*, 2003. 5(2): pp. 70–73.

[8] Lovelock, J.E., Notarized suggestion from 1958, Cited in: Lee M.L., Markides K.E., eds., Provo, UT: Chromatography Conference, Inc. Analytical Supercritical Fluid Chromatography and Extraction, 1990: pp. 7.

[9] Schoenmakers, P.J., Chapter 4: Open column or packed columns for supercritical fluid chromatography, in *Supercritical Fluid chromatography RM Smith ed. RSC Chromatography Monographs.* The Royal Society of Chemistry, London, 1988.

[10] Berger, T.A. and W.H. Wilson, Packed column supercritical fluid chromatography with 220,000 plates. *Analytical Chemistry,* 1993. 65(10): pp. 1451–1455.

[11] Cui, Y. and S.V. Olesik, High-performance liquid chromatography using mobile phases with enhanced fluidity. *Analytical Chemistry*, 1991. 63(17): pp. 1812–1819.

[12] Searle, P.A., K.A. Glass, and J.E. Hochlowski, Comparison of preparative HPLC/MS and preparative SFC techniques for the high-throughput purification of compound libraries. *Journal of Combinatorial Chemistry*, 2004. 6(2): pp. 175–180.

[13] Desfontaine, V., et al., Supercritical fluid chromatography in pharmaceutical analysis. *Journal of Pharmaceutical and Biomedical Analysis,* 2015. 113: pp. 56–71.

[14] Schafer, W., et al., Improved chiral SFC screening for analytical method development. *Chirality*, 2013. 25(11): pp. 799–804.

[15] Barhate, C.L. et al., Ultrafast chiral separations for high throughput enantiopurity analysis. *Chemical Communications,* 2017. 53(3): pp. 509–512.

[16] Kalíková, K. et al., Supercritical fluid chromatography as a tool for enantioselective separation; A review. *Analytica Chimica Acta*, 2014. 821(Supplement C): pp. 1–33.

[17] Barhate, C.L. et al., Instrumental idiosyncrasies affecting the performance of ultrafast chiral and achiral sub/supercritical fluid chromatography. *Analytical Chemistry*, 2016. 88(17): pp. 8664–8672.

[18] Nováková, L., et al., Modern analytical supercritical fluid chromatography using columns packed with sub-2 μm particles: A tutorial. *Analytica Chimica Acta*, 2014. 824: pp. 18–35.

[19] Liu, Y. et al., Super/subcritical fluid chromatography chiral separations with macrocyclic glycopeptide stationary phases. *Journal of Chromatography A*, 2002. 978(1): pp. 185–204.

[20] Barhate, C.L., et al., High efficiency, narrow particle size distribution, sub-2 μm based macrocyclic glycopeptide chiral stationary phases in HPLC and SFC. *Analytica Chimica Acta*, 2015. 898(Supplement C): pp. 128–137.

[21] Berger, T.A., Kinetic performance of a 50mm long 1.8 μm chiral column in supercritical fluid chromatography. *Journal of Chromatography A*, 2016. 1459: pp. 136–144.

[22] Dispas, A., et al., Evaluation of the quantitative performances of supercritical fluid chromatography: From method development to validation. *Journal of Chromatography A*, 2014. 1353(Supplement C): pp. 78–88.

[23] Wang, Z., et al., Development of an orthogonal method for mometasone furoate impurity analysis using supercritical fluid chromatography. *Journal of Chromatography A*, 2011. 1218(16): pp. 2311–2319.

[24] Regalado, E.L., et al., Chromatographic resolution of closely related species: drug metabolites and analogs. *Journal of Separation Science*, 2014. 37(9–10): pp. 1094–1102.

[25] Barhate, C.L., et al., Ultrafast separation of fluorinated and desfluorinated pharmaceuticals using highly efficient and selective chiral selectors bonded to superficially porous particles. *Journal of Chromatography A*, 2015. 1426: pp. 241–247.

[26] Regalado, E.L., et al., Chromatographic resolution of closely related species in pharmaceutical chemistry: dehalogenation impurities and mixtures of halogen isomers. *Analytical Chemistry*, 2013. 86(1): pp. 805–813.

[27] Venkatramani, C., et al., Simultaneous achiral-chiral analysis of pharmaceutical compounds using two-dimensional reversed phase liquid chromatography-supercritical fluid chromatography. *Talanta*, 2016. 148: pp. 548–555.

[28] Berger, T.A., Instrumentation for analytical scale supercritical fluid chromatography. *Journal of Chromatography A*, 2015. 1421(Supplement C): pp. 171–183.

[29] Maftouh, M., et al., Screening approach for chiral separation of pharmaceuticals Part III. Supercritical fluid chromatography for analysis and purification in drug discovery. *Journal of Chromatography A*, 2005. 1088(1–2): pp. 67–81.

[30] Berthod. A. ed. *Chiral Recognition in Separation Methods. Mechanisms and Applications*, 2010. Berlin, Heidelberg: Springer-Verlag.

[31] Armstrong, D.W., Optical isomer separation by liquid chromatography. *Analytical Chemistry*, 1987. 59(2): pp. 84A–91A.

[32] Pirkle, W.H. and T.C. Pochapsky, Considerations of chiral recognition relevant to the liquid chromatography separation of enantiomers. *Chemical Reviews*, 1989. 89(2): pp. 347–362.

[33] Taylor, D.R. and K. Maher, Chiral separations by high-performance liquid chromatography. *Journal of Chromatographic Science*, 1992. 30(3): pp. 67–85.

[34] Okamoto, Y. and T. Ikai, Chiral HPLC for efficient resolution of enantiomers. *Chemical Society Reviews*, 2008. 37(12): pp. 2593–2608.

[35] Anton, K., *Supercritical Fluid Chromatography with Packed Columns: Techniques and Applications* 1997, New York: Marcel Dekker..

[36] Berger, T.A., High-efficiency packed-column supercritical-fluid chromatography of sulfonylurea herbicides and metabolites from large water samples. *Chromatographia*, 1995. 41(3–4): pp. 133–140.

[37] Berger, T.A. and W.H. Wilson, Separation of basic drugs by packed-column supercritical-fluid chromatography.3. Stimulants. *Journal of Pharmaceutical Sciences*, 1995. 84(4): pp. 489–492.

[38] Berger, T.A. and R.M. Smith. Packed Column SFC. 2007; Available from: https://nls.ldls.org.uk/welcome.html?ark:/81055/vdc_100035466745.0x000001.

[39] Anton, K., et al., Chiral separations by packed-column super-critical and subcritical fluid chromatography. *Journal of Chromatography A*, 1994. 666(1–2): pp. 395–401.

[40] Inc, C.T., *Technical Support, Products and Services for Chiral Analysis and Separation*, 2004.

[41] Khater, S. and C. West, Insights into chiral recognition mechanisms in supercritical fluid chromatography V. Effect of the nature and proportion of alcohol mobile phase modifier with amylose and cellulose tris-(3,5-dimethylphenylcarbamate) stationary phases. *Journal of Chromatography A*, 2014. 1373: pp. 197–210.

[42] Okamoto, Y., M. Kawashima, and K. Hatada, Chromatographic resolution. 7. Useful chiral packing materials for high-performance liquid chromatographic resolution of enantiomers: phenylcarbamates of polysaccharides coated on silica gel. *Journal of the American Chemical Society*, 1984. 106(18): pp. 5357–5359.

[43] Okamoto, Y., M. Kawashima, and K. Hatada, Chromatographic resolution: XI. Controlled chiral recognition of cellulose triphenylcarbamate derivatives supported on silica gel. *Journal of Chromatography A*, 1986. 363(2): pp. 173–186.

[44] Okamoto, Y., et al., Useful chiral stationary phases for HPLC. Amylose Tris(3,5–dimethylphenylcarbamate) and Tris(3,5-dichlorophenylcarbamate) supported on silica gel. *Chemistry Letters*, 1987. 16(9): pp. 1857–1860.

[45] Okamoto, Y. and Y. Kaida, Resolution by high-performance liquid chromatography using polysaccharide carbamates and benzoates as chiral stationary phases. *Journal of Chromatography A*, 1994. 666(1): pp. 403–419.

[46] Zhang, T., et al., Cellulose tris(3,5-dichlorophenylcarbamate) immobilised on silica: A novel chiral stationary phase for resolution of enantiomers. *Journal of Pharmaceutical and Biomedical Analysis*, 2008. 46(5): pp. 882–891.

[47] De Klerck, K., Y. Vander Heyden, and D. Mangelings, Pharmaceutical-enantiomers resolution using immobilized polysaccharide-based chiral stationary phases in supercritical fluid chromatography. *Journal of Chromatography A*, 2014. 1328: pp. 85–97.

[48] Franco, P. and T. Zhang, Common approaches for efficient method development with immobilised polysaccharide-derived chiral stationary phases. *Journal of Chromatography B*, 2008. 875(1): pp. 48–56.

[49] Chankvetadze, B., Recent developments on polysaccharide-based chiral stationary phases for liquid-phase separation of enantiomers. *Journal of Chromatography A*, 2012. 1269: pp. 26–51.

[50] Zhang, T., D. Nguyen, and P. Franco, Enantiomer resolution screening strategy using multiple immobilised polysaccharide-based chiral stationary phases. *Journal of Chromatography A*, 2008. 1191(1): pp. 214–222.

[51] Zhang, T. et al., Solvent versatility of immobilized 3,5-dimethylphenylcarbamate of amylose in enantiomeric separations by HPLC. *Journal of Chromatography A*, 2005. 1075(1): pp. 65–75.

[52] Bargmann-Leyder, N. et al., Evaluation of Pirkle-type chiral stationary phases by liquid and supercritical fluid chromatography: Influence of the spacer length and the steric hindrance in the vicinity of the stereogenic centre. *Journal of Chromatography A*, 1994. 666(1): pp. 27–40.

[53] Pirkle, W.H., D.W. House, and J.M. Finn, Broad spectrum resolution of optical isomers using chiral high-performance liquid chromatographic bonded phases. *Journal of Chromatography A*, 1980. 192(1): pp. 143–158.

[54] Williams, K.L. and L.C. Sander, Enantiomer separations on chiral stationary phases in supercritical fluid chromatography. *Journal of Chromatography A*, 1997. 785(1): pp. 149–158.

[55] Terfloth, G.J., et al., Broadly applicable polysiloxane-based chiral stationary-phase for high-performance liquid-chromatography and supercritical-fluid chromatography. *Journal of Chromatography A*, 1995. 705(2): pp. 185–194.

[56] Pirkle, W.H. and G.J. Terfloth, Naproxen-derived segmented and side-chain-modified polysiloxanes as chiral stationary phases. *Journal of Chromatography A*, 1995. 704(2): pp. 269–277.

[57] Curran, D.P., Fluorous reverse phase silica gel. A new tool for preparative separations in synthetic organic and organofluorine chemistry. *Synlett*, 2001. 9: pp.1488–1496.

[58] Armstrong, D.W., et al., Macrocyclic antibiotics as a new class of chiral selectors for liquid chromatography. *Analytical Chemistry*, 1994. 66(9): pp. 1473–1484.

[59] Barhate, C.L., et al., High efficiency, narrow particle size distribution, sub-2 μm based macrocyclic glycopeptide chiral stationary phases in HPLC and SFC. *Analytica Chimica Acta*, 2015. 898: pp. 128–137.

[60] Moskaľová, M., et al., HPLC Enantioseparation of novel spirobrassinin analogs on the cyclofructan chiral stationary phases. *Chromatographia*, 2017. 80(1): pp. 53–62.

[61] Phillips, J.H. and R. J. Robey, Solvent strength and selectivity properties of supercritical carbon dioxide relative to liquid hexane. *Journal of Chromatography A*, 1989. 465(3): pp. 177–188.

[62] Klesper, E. and W. Hartmann, Parameters in supercritical fluid chromatography of styrene oligomers. *Journal of Polymer Science: Polymer Letters Edition*, 1977. 15(12): pp. 707–712.

[63] Lesellier, E. and C. West, The many faces of packed column supercritical fluid chromatography – A critical review. *Journal of Chromatography A*, 2015. 1382: pp. 2–46.

[64] Berger, T.A., Separation of polar solutes by packed column supercritical fluid chromatography. *Journal of Chromatography A*, 1997. 785(1): pp. 3–33.

[65] Lesellier, E. and C. West, The many faces of packed column supercritical fluid chromatography – A critical review. *Journal of Chromatography A*, 2015. 1382: pp. 2–46.

[66] Taylor, L.T., Packed column supercritical fluid chromatography of hydrophilic analytes via water-rich modifiers. *Journal of Chromatography A*, 2012. 1250: pp. 196–204.

[67] Tarafder, A., Metamorphosis of supercritical fluid chromatography to SFC: An Overview. *Trac-Trends in An alytical Chemistry*, 2016. 81: pp. 3–10.

[68] Swann, T., et al. Using blends of solvents and additives to enhance SFC chiral method development screening. 2015; available from: http://www.waters.com/webassets/cms/library/docs/2015sfc_swann_trefoilmethdev.pdf.

[69] DaSilva, J.O. et al., Evaluation of non-conventional polar modifiers on immobilized chiral stationary phases for improved resolution of enantiomers by supercritical fluid chromatography. *Journal of Chromatography A*, 2014. 1328: pp. 98–103.

[70] Miller, L., Use of dichloromethane for preparative supercritical fluid chromatographic enantioseparations. *Journal of Chromatography A*, 2014. 1363: pp. 323–330.

[71] Lee, J., et al., On the method development of immobilized polysaccharide chiral stationary phases in supercritical fluid chromatography using an extended range of modifiers. *Journal of Chromatography A*, 2014. 1374: pp. 238–246.

[72] Taylor, L.T., Supercritical fluid chromatography. *Analytical Chemistry*, 2008. 80(12): p. 4285–4294.

[73] Phinney, K.W. and L.C. Sander, Additive concentration effects on enantioselective separations in supercritical fluid chromatography. *Chirality*, 2003. 15(4): pp. 287–294.

[74] Ye, Y.K., et al., Enantioseparation of amino acids on a polysaccharide-based chiral stationary phase. *Journal of Chromatography A*, 2002. 945(1–2): pp. 147–159.

[75] Barnhart, W.W., et al., Supercritical fluid chromatography tandem-column method development in pharmaceutical sciences for a mixture of four stereoisomers. *Journal of Separation Science*, 2005. 28(7): pp. 619–626.

[76] Welch, C.J., et al., Solving multicomponent chiral separation challenges using a new SFC tandem column screening tool. *Chirality*, 2007. 19(3): pp. 184–189.

[77] Wang, C., A.A. Tymiak, and Y. Zhang, Optimization and simulation of tandem column supercritical fluid chromatography separations using column back pressure as a unique parameter. *Analytical Chemistry*, 2014. 86(8): pp. 4033–4040.

[78] Wang, C., A.A. Tymiak, and Y. Zhang, Chapter 6 – Application of multiple column supercritical fluid chromatography A2. In *Supercritical Fluid Chromatography*, Colin F. Poole (ed.), 2017, Elsevier. pp. 153–172.

[79] Delahaye, S. and F.d.r. Lynen, Implementing stationary-phase optimized selectivity in supercritical fluid chromatography. *Analytical Chemistry*, 2014. 86(24): pp. 12220–12228.

[80] Searle, P.A. and E.E. Jordan, Strategies for increasing throughput of chiral separations by supercritical fluid chromatography. *Chromatography Today*, 2015. August.

[81] Searle, P.A. and E.E. Jordan, Supercritical fluid chromatography: an essential tool in drug discovery. *American Pharmaceutical Review*, 2017. March.

Index

https://doi.org/10.1515/9783110500776-009